Fundamentals of Computer Networks

Matthew N. O. Sadiku
Cajetan M. Akujuobi

Fundamentals of Computer Networks

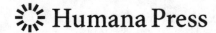

Humana Press

Matthew N. O. Sadiku
Prairie View A&M University
Prairie View, TX, USA

Cajetan M. Akujuobi
Prairie View A&M University
Prairie View, TX, USA

ISBN 978-3-031-09419-4   ISBN 978-3-031-09417-0   (eBook)
https://doi.org/10.1007/978-3-031-09417-0

This Springer imprint is published by the registered company Springer Nature Switzerland AG
The registered company address is: Gewerbestrasse 11, 6330 Cham, Switzerland

Dedicated to our wives:
Janet and Caroline

Preface

Computer networking is an exciting and fascinating field of study. It is the fastest growing technology in our society. A computer network is a connection of two or more autonomous computers and other devices so that they can communicate and share information. It enables sharing of files, data, and devices such as printers. It facilitates interpersonal communications and allows users to communicate via various means: email, instant messaging, chat rooms, telephone, video telephone calls, and video conferencing. Computer networking relies on the related discipline of electrical engineering, computer engineering, and computer science.

Earlier, computer networks consisted of mainframes in a building. Computers became more powerful, smaller, and less expensive, to the point that the typical desktop PC is equivalent to an early mainframe. This led to the need to network those devices within the workplace to form a local area network (LAN) that will allow PC users to share access to hosts, databases, printers, etc.

The growth of computer networks from local area networks (LANs) to the Internet has been rapid. Right now, we have many different types of computer networks. They are often classified in terms of their geographical coverage:

- Local area network (LAN), which covers a building or campus
- Metropolitan area network (MAN), which covers a city
- Wide area network (WAN), which can cover a nation, continent, or anywhere on Earth

The most popular computer network is the Internet, which is a WAN. Global Internet enables its users to access enormous amount of information worldwide.

Computer networks have emerged as the integration of technologies in computer and communication, and their evolution is tightly related to the advances in these two areas. Our modern society has been greatly affected by computer networks, especially the Internet. Computer networks have changed how we live, educate, entertain, and do business. Within a short time, computer communication networks have become an indispensable part of business, industry, and entertainment. Industries, government agencies, military, education, libraries, healthcare, law enforcement, justice, manufacturing, financial services, insurance, transportation, aerospace, telecommunication, energy, retail, and utilities are becoming increasingly dependent on the computer networks. People now can easily and conveniently connect to any-body anywhere in the world, send emails, access information, do shopping, and chat on the Internet without leaving their homes.

The major objectives of this book are:

- To introduce students to how computer networks work
- To familiarize students with international standards for computer networks
- To bring to students' awareness the newly emerging technologies in computer networking

Most books on computer networks are designed for a two-semester course sequence. Unfortunately, electrical and computer engineering has grown considerably, and its curriculum is so crowded that there is no room for a two-semester course on computer networks. This book is designed for a 3-hour semester course on computer networks. It is intended as a textbook for senior-level students in electrical and computer engineering. Their mathematical background should include calculus and probability. But no advanced mathematics is required.

This book is intended to present computer networks to electrical and computer engineering students in a manner that is clearer, more interesting, and easier to understand than other texts. All principles are presented in a lucid, logical, and step-by-step manner. As much as possible, we avoid wordiness and giving too much detail that could hide concepts and impede overall understanding of the material. Ten review questions in the form of multiple-choice objective items are provided at the end of each chapter with answers. The review questions are intended to cover the little "tricks" which the examples and end-of-chapter problems may not cover. They serve as a self-test device and help students determine how well they have mastered the chapter. A problem solution manual will be provided, and it will be available directly from the publisher.

The book is organized to serve as a textbook for an undergraduate course in computer networks. The book is divided into 12 chapters, with each chapter self-contained.

Chapter 1: "Introduction." The introductory chapter defines computer networks and provides examples. It covers common protocols, types of computer network (LANs, MANs, and WANs), and popular applications of the computer networks.

Chapter 2: "Digital Communications." The idea of a digital communication system and all of the various components that make up the entire system are discussed in this chapter. This includes the transmission media, the encoding techniques, bit/byte stuffing, multiplexing, and the different types of switching techniques.

Chapter 3: "Network Models." In this chapter, the network models such as the Open Systems Interconnection (OSI) and the Transmission Control Protocol/Internet Protocol (TCP/IP) models are presented. This chapter also discusses the different networking applications—the internetworking system devices and Signaling Systems No. 7 (SS7).

Chapter 4: "Local Area Networks." This chapter covers the local area networks (LANs) and their different types. It mentions the advantages and disadvantages of the different topologies of LAN. All of the operational access methodologies including the controlled access devices are also discussed.

Chapter 5: "The Internet." This chapter introduces TCP/IP protocols and IP addresses. It discusses important issues facing the Internet: privacy, security, and safety. It also covers the next-generation Internet (IPv6) and Internet2, which is the future of Internet.

Chapter 6: "Intranets and Extranets." The excitement created by the Internet (a public network) has been transferred to modern networks called intranets and extranet. This chapter presents an introduction to intranets and extranets as two of the growing applications of Internet.

Chapter 7: "Virtual Private Networks." This chapter covers the main characteristics of virtual private networks (VPN), different types of VPNs, various applications of VPNs, and the benefits and challenges of VPNs.

Chapter 8: "Digital Subscriber Line." The chapter discusses digital subscriber line (DSL). DSL is a high-speed data technology which is also referred to as a broadband telecommunication system and constitutes part of the access technology network. The comparison between the different types of DSL is also made.

Chapter 9: "Optical Networks." This chapter gives an introduction to optical networks. It discusses some of the main optical components. It covers various optical networks: WDM-based networks, passive optical networks, SONET, all-optical networks, and free-space optics. It provides some applications of optical networks.

Chapter 10: "Wireless Networks." This chapter provides a brief overview of wireless communication networks. It considers wireless local area networks (WLANs), wireless metropolitan area networks (WMANs), wireless wide area network (WWAN), and wireless personal area network (WPAN). It explains cellular network, satellite networks, and wireless sensor network (WSN).

Chapter 11: "Network Security." In this chapter, the important aspects of network security are discussed. The chapter describes malware, firewall, encryption, digital signatures, intrusion detection and prevention, and cybersecurity and how they relate to network security.

Chapter 12: "Emerging Technologies." This chapter discusses some of the major emerging technologies including Internet of Things (IoT), big data, smart cities, blockchain technology, cloud computing, fog computing, edge computing, 5G networks, and cybersecurity issues. Steganography, an information security technique used in cybersecurity related issues, is also covered.

Appendix A covers old technologies such as X.25, Frame Relay, ISDN and BISDN, ATM, and MPLS, while Appendix B gives a short introduction to queuing theory.

   This book provides a comprehensive introduction to the key concepts of computer networks in a manner that is easily digestible for a beginner in the field or undergraduate students. The textbook is primarily intended for major universities and colleges offering courses on computer networks. It is written for a one-semester senior-level course on computer networks. It may also serve as a reference or a quick review of fundamentals of computer networks to people in industry. A practicing engineer who needs an overview of computer networks can also benefit. We are grateful for the support of Dr. Annamalia Annamalai, the department head of the Department of Electrical and Computer Engineering, and Dr. Pamela Obiomon, the dean of the College of Engineering at Prairie View A& M University, Prairie View, Texas.

Prairie View, TX, USA   Matthew N. O. Sadiku
                                                             Cajetan M. Akujuobi

Contents

About the Authors

Matthew N. O. Sadiku He received his BSc degree in 1978 from Ahmadu Bello University, Zaria, Nigeria, and his MSc and PhD degrees from Tennessee Technological University, Cookeville, TN, in 1982 and 1984, respectively. From 1984 to 1988, he was an assistant professor at Florida Atlantic University, Boca Raton, FL, where he did graduate work in computer science. From 1988 to 2000, he was at Temple University, Philadelphia, PA, where he became a full professor. From 2000 to 2002, he was with Lucent/Avaya, Holmdel, NJ, as a system engineer and with Boeing Satellite Systems, Los Angeles, CA, as a senior scientist. He is presently Regents Professor Emeritus of electrical and computer engineering at Prairie View A&M University, Prairie View, TX.Mathew is the author of over 1,010 professional papers and over 100 books including *Elements of Electromagnetics* (Oxford University Press, 7th ed., 2018), *Fundamentals of Electric Circuits* (McGraw-Hill, 7th ed., 2020, with C. Alexander), *Computational Electromagnetics with MATLAB* (CRC Press, 4th ed., 2019), *Principles of Modern Communication Systems* (Cambridge University Press, 2017, with S. O. Agbo), and *Emerging Internet-based Technologies* (CRC Press, 2019). In addition to the engineering books, he has written Christian books including *Secrets of Successful Marriages* and *How to Discover God's Will for Your Life*, as well as commentaries on all the books of the New Testament Bible. Some of his books have been translated into French, Korean, Chinese (and Chinese Long Form in Taiwan), Italian, Portuguese, and Spanish.Mathew was the recipient of the 2000 McGraw-Hill/Jacob Millman Award for outstanding contributions in the field of electrical engineering. He was also the recipient of regents professor award for 2012–2013 by the Texas A&M University System. He is a registered professional engineer and a life fellow of the Institute of Electrical and Electronics Engineers (IEEE) "for contributions to computational electromagnetics and engineering education." He was the IEEE Region 2 Student Activities Committee Chairman. Mathew was an associate editor for *IEEE Transactions on Education.* He is also a member of Association for Computing Machinery (ACM) and American Society of Engineering Education (ASEE). His current research interests are in the areas of computational electromagnetic, computer networks, and engineering education. His works can be found in his autobiography, *My Life and Work* (Trafford Publishing, 2017) or his website: www.matthew-sadiku.com. He currently resides with his wife Janet in West Palm Beach, Florida. He can be reached via email at sadiku@ieee.org

Cajetan M. Akujuobi He received his OND from the Institute of Management and Technology, Enugu, Nigeria in 1974; his BS degree from Southern University, Baton Rouge, Louisiana, in 1980; his MS degree from Tuskegee University, Alabama in 1983, all in electrical & electronics engineering; his PhD degree from George Mason University, Fairfax, Virginia in 1995, in electrical engineering; and his MBA degree from Hampton University. He is also a licensed professional engineer (PE) in the State of Texas.He is a full professor in the Department of Electrical & Computer Engineering and was the former vice president for research, innovation, and sponsored programs at Prairie View A&M University (PVAMU), a position he held from January 2, 2014, to August 31, 2018. During that period, he was instrumental in bringing to PVAMU five new Chancellors' Research Initiative (CRI) research centers worth over $35 Million from 2014 to 2018. Prof. Akujuobi grew the research expenditure of the University by over 10% annually. He also served as the Dean for Graduate Studies at PVAMU for over 3 years. He is the founding executive director of many research programs including the Center of Excellence for Communication Systems Technology Research (CECSTR), a Texas A&M Board of Regents approved center, where he was able to attract research funding exceeding over $25 Million. He is also the principal investigator and founder of the Cybersecurity Center of Excellence (SECURE: Systems to Enhance CYBERSECURITY for Universal Research Environment) where he received over $7 Million research awards in 2016 as the principal investigator.Prof. Akujuobi was the founding dean of the College of Science, Technology, Engineering, and Mathematics at Alabama State University (ASU) from June 12, 2010, to December 31, 2014. At ASU, he founded two new research centers, and was founding executive director of the Center of Excellence for Communication Systems & Image/Signal/Video Processing (CECSIP) and became the founding executive director of the STEM Center of Excellence for Modeling & Simulation Research (SCEMSR). Under his leadership, ASU got approval to start their forensic biology program and master's course in forensic science, and he was instrumental in introducing the biomedical engineering program as well. Research and enrollment grew in his college by more than 5% annually during his tenure at ASU.He has published extensively including writing books and book chapters. Two of the books he published in 2008 with M. N. O. Sadiku are *Introduction to Broadband Communication Systems* and *Solution Manual for Introduction to Broadband Communication Systems*, both published by Chapman & Hall/CRC and Sci-Tech.Prof. Akujuobi is the current chair of the IEEE Houston Section Life Members Group. He is also a life senior member of IEEE, senior member of Instrument Society of America (ISA), and member of the American Society for Engineering Education (ASEE), Sigma XI, the Scientific Research Society, and the Texas Society for Biomedical Research (TSBR) Board of Directors. He was selected as one of the US representatives for engineering educational and consultation mission to Asia in 1989. He is listed in *Who's Who in Science and Engineering*, *Who's Who in the World*, *Who's Who in America*, *Who's Who in American Education*, and *Who's Who in Industry & Finance*.

Introduction

1

Even though most people won't be directly involved with programming, everyone is affected by computers, so an educated person should have a good understanding of how computer hardware, software, and networks operate.

Brian Kernighan

Abstract

This chapter introduces the reader to what computer networking is all about. Computer networks differ in the transmission medium (wired or wireless), topology, size, and scope. They may be classified as switched or dedicated. They may also be classified as local area networks, metropolitan area networks, and wide area networks. The chapter presents common applications of computer networks.

Keywords

Computer networks · Local area networks · Metropolitan area networks · Wide area networks

1.1 What Are Computer Networks?

Computer networking is an exciting and fascinating field of study. It is the fastest growing technology in our society. A computer network may be regarded as part of the general area of communication networks. Communication networks have evolved from the telegraph to computer nets:

(a) Telegraph network: It uses Morse code to send messages between users.
(b) Telephone network: It uses circuit switching to exchange messages between two or more users.
(c) A computer network is an interconnection of autonomous computers or digital devices.

A **computer network** is an interconnection of computing devices that facilitates communication among the devices.

A computer network is used in:

- Providing a transfer of information between users. It facilitates interpersonal communication via email, video conferencing, instant messaging, etc.
- Providing users with access to databases
- Sharing expensive resources such as color printer or scanner among multiple users
- Sharing file or software between hosts across the network

To set up a computer network may require these items: network cables, routers, network

M. N. O. Sadiku, C. M. Akujuobi, *Fundamentals of Computer Networks*,
https://doi.org/10.1007/978-3-031-09417-0_1

cards, software, and computers/workstations/ devices to be connected. The software consists of network operating system (NOS) and specific application programs. The NOS coordinates multiple computers across a network and keeps the network running smoothly. Examples of NOSs are AppleShare, LANtastic, and Novell NetWare.

For wireless network, network cards are necessary, but no cable is needed.

In order for the devices connected to a computer network to communicate with each other, they must speak the same language, i.e., a common language, called *protocol*, must be established.

> A **protocol** is a set of rules and procedures that permit orderly exchange of information within a network.

Computer networks extensively use protocols to accomplish various communication tasks. Two examples of popular protocols are Real-time Transport Protocol (RTP) which supports real-time services such as audio and video and Internet Protocol Security (IPSec) which is regarded as the "standard" for achieving secure communications over the Internet.

Communication networks can be classified as switched or dedicated. A switched network is composed of three phases—call setup, call transmission, and call termination. A dedicated network consists of private lines for businesses that

need 24-hour-a-day access. In either case, a communication network consists of a transmitter, a receiver, and a transmission medium, which may be wired or wireless. A computer network consists of a set of digital devices using common protocols to communicate over connecting transmission media.

Computer networks differ in the transmission medium (wired or wireless), topology, size, and scope. In terms of their geographical scope, computer networks may be classified as:

(a) Local area network (LAN) covering 0–2 km, e.g., within a room, a single building, and a campus. Typical examples are the Ethernet and Cambridge ring. LANs have low delay and make few errors. They operate on moderately high data rates (e.g., 1–10 Mbps). Typically, Ethernet operates at data transfer rates of 10 Mbps to 400 Gbps. A typical LAN is shown in Fig. 1.1.

(b) Metropolitan area network (MAN) covers a city, extending from 2 to 50 km. Typical MANs include FDDI (Fiber Distributed Data Interface) and WiMAX (Worldwide Interoperability for Microwave Access). The cable TV network is another example. A MAN is basically a high-speed, backbone network that interconnects a number of LANs in a large city or metropolitan area. It occupies a middle ground between LANs and WANs.

(c) Wide area network (WAN) (or long-haul) covers a nation, a continent, or the entire world. Examples of WANs are ARPANET, TYMNET, and the Internet (the so-called

Fig. 1.1 A simple LAN. (Source: "LANs and WANs Explained," by Robert Hopkins, 2017, http://eas-tech.net/ lans-wans-explained)

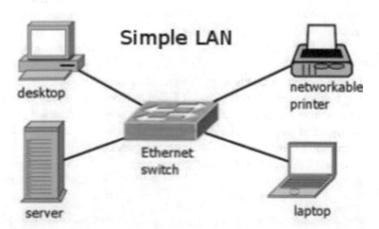

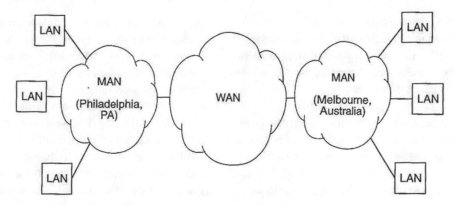

Fig. 1.2 Interconnection of LANs, MANs, and WANs. (Source: Sadiku 1995, p. 2)

Information Superhighway). The telephone network is another example of a WAN.

Since WAN can cover the globe, its use is being addressed by various national initiatives. A relationship between LANs, MANs, and WANs is shown in Fig. 1.2. Other types of computer networks include body area network (BAN), personal area network (PAN), home area network (HAN), storage area network (SAN), campus area network (CAN), and global area network (GAN).

1.2   Short History

Early computer networks were research networks and were usually administered by government agencies and supported by government grants. Significant contributions to the development of computer networks include:

- Digital computers were born in the 1940s and started multiplying in the 1960s.
- Before 1950, there were telegraph and telephone networks.
- In the late 1950s, a network of communicating computers functioned as part of the US military's Semi-Automatic Ground Environment (SAGE) radar system.
- In 1969, UCLA, the Stanford Research Institute, the University of California at Santa Barbara, and the University of Utah were connected as the beginning of the Advanced Research

Projects Agency Network (ARPANET) . ARPANET later evolved into the Internet.
- ALOHA network, a satellite network, was developed in Hawaii.
- In 1972, commercial services using X.25 were deployed.
- In 1973, Robert Metcalfe wrote a formal memo at Xerox PARC describing Ethernet. He was inspired by the ALOHA network, a packet radio network developed at the University of Hawaii.
- In 1974, IBM introduced its Systems Network Architecture (SAN), while Digital Equipment introduced its DECnet in 1975.
- In 1976, John Murphy of Datapoint Corporation created ARCNET, a token-passing network first used to share storage devices.
- In 1981, Computer Science Network (CSNET) was sponsored by NSF and created for universities that could not use ARPANET.
- The MAN effort started in 1982, about 2 years after IEEE Project 802 was initiated.
- The 1990s witnessed the emergence of the World Wide Web, which enables browsing and allows access to all kinds of information on the Internet.
- In 1991, three companies (IBM, Merit, and Verizon) built a high-speed network called Advanced Network Services Network (ANSNET).
- In April 1995, Netscape became the most popular navigator. In August 1995, Microsoft released the first version of Internet Explorer.

- In 1995, the transmission speed capacity for Ethernet increased from 10 Mbps to 100 Mbps. By 1998, Ethernet supported transmission speeds of a Gigabit. Currently, 400 Gbps Ethernet is being developed.

Computer networks have literally grown exponentially and become an integral part of our modern life. The growth is interesting and exciting. It can be attributed to the expansion of Internet access to almost every section of our society.

1.3　Applications

We live in an exciting time in which we are directly or indirectly involved in computer communication networks such as Internet. Businesses of every size and kind, education systems, government at all levels, entertainment industries, and organizations embrace computer networks as a means of sharing information and resources, conducting commerce, and engaging socially. Here we discuss some of the common applications of computer networks.

- *Electronic mail*: This the most common use of computer networks. It allows a user to send a message to another user. Closely related to this is *file transfer* (also known as FTP). It allows files to be transferred from one computer to another.
- *Government:* Computer networks are making significant changes in all levels of government—federal, state, and local. Digital government (or networked-government) refers to the use of information technologies such as the Internet and mobile computing to support government operations and provide government services. It has improved the access of citizens to services. It has made government to be more effective, accessible, and transparent. Digital government enables citizens to interact with the government and improves the services that governments offer their citizens. It is already making a positive impact on the government, citizens, business, and society. A noticeable impact is paperless electronic voting, which outperforms traditional voting in many ways. These include accessibility and easy ballot management. Electronic voting can lead to efficient voting and smarter government.
- *Ecommerce*: Electronic commerce is the process of conducting business over computer networks such as the Internet. It has revolutionized business transactions by enabling consumers to purchase, invest, bank, and communicate from virtually anytime, anywhere. It enables such transactions as online shopping, online banking, payment systems, electronic air tickets, hotel reservation, tourism, and teleconferencing. The ecommerce has become an indispensable tool for businesses worldwide because it allows people to conduct business transactions and reach consumers directly.
- *Education*: Computer networks such as the Internet have transformed education at all levels to meet the demand of the twenty-first century. Online education has a great potential to reach students with personalized education at a low cost. It is gaining ground as an extension of traditional education. The emergence of social networking technologies is rapidly changing the delivery of online education. By taking advantage of these technologies, online education can provide quality education anywhere, anytime. Education providers are moving from traditional face-to-face environments to those that are completely electronic. Online education provides university equivalent courses for millions of students across the globe.
- *Social media*: When computers connect people with machines, they become social networks. Social networks have now become popular means of social networking and communication about social events. They facilitate interaction with family and friends who share their personal or political interests. They allow the discussion of self-actualization, self-presentation, and self-disclosure of individuals. They have become a part of our routines and are widely accepted by digital citizens. Although there are several social network sites such as Google, Twitter, Myspace, and

LinkedIn, Facebook is the most popular. The amount of sharing and cooperation that occurs in online communities is the key advantage.

Other applications of computer networks include manufacturing, digital games, digital library, electronic health, mobile banking, online religion, smart grid, smart agriculture, and smart cities.

1.4   Standards

Standards play an important role in engineering in general and computer networks in particular. Without standards, vendors will manufacture products that will not interoperate with products from other vendors. Standards take years to develop and ratify; it is a gradual process. They are created by bringing together experience and expertise of stakeholders on specific issues. For a specification to become a standard, it must be well-understood and thoroughly tested, have interoperable implementations, and enjoy public support.

> A **standard** is a thoroughly tested specification that has been approved by a standards organization.

There are several organizations that develop standards:
(a) International Standards Organization (ISO)
(b) Institute of Electrical and Electronics Engineers (IEEE)
(c) International Telecommunications Union (ITU)
(d) Internet Engineering Task Force (IETF)
(e) European Telecommunications Standards Institute (ETSI)
(f) American National Standard Institute (ANSI)
(g) National Institute of Standards and Technology (NIST)
(h) Japanese Telecommunications Technology Committee (TTC)

(i) Telecommunications Standards Advisory Council of Canada (TSACC)

Different countries and regions often have their own standards.

Of particular importance are ITU and IEEE standards. ITU is a branch of the United Nations. It allocates spectrum for global and satellite communications. It develops technical standards that ensure that network products seamlessly interconnect.

IEEE is perhaps the largest technical professional organization. It publishes technical information in the form of papers, sponsors conferences, and develops standards. IEEE 802 is a family of IEEE standards dealing with LANs and MANs. IEEE 802.3 is for Ethernet, which transmits data over both copper and fiber cables. IEEE 803.11 is for the standard for wireless LAN (also known Wi-Fi). It is perhaps the most well-known member of the IEEE 802 family for home users.

Besides IEEE, the Internet Engineering Task Force (IETF) (established by the IAB, Internet Activities Board) published a series of documents known as Requests for Comments (RFCs). The RFCs are numbered sequentially in chronological order. They are submitted by experts and may later become standards. An RFC usually progresses through three stages: proposed standard, draft standard, and Internet standard. Simple Network Management Protocol (SNMP) is a typical example of an RFC that went through the procedure.

Summary
1. A computer network consists of a set of digital devices using common protocols to communicate over connecting transmission media.
2. Computer networks are usually classified as local area networks (LANs), metropolitan area networks (MANs), and wide area networks (WANs).
3. Early computer networks include ARPANET, X.25, and Ethernet.
4. Computer networks support an enormous number of applications and services such as ecommerce, ehealth, online education, online

banking, social media, smart grid, and open government.

5. A standard is a thoroughly tested specification that has been approved by a standards organization. Standards organization include IEEE, ITU, ISO, and ANSI.

Review Questions

1.1 A network that requires call setup, call transmission, and call termination is known as:
(a) Dedicated network   (b) Switched network   (c) Long-haul   network   (d) Local network

1.2 Which of the following computer network typically has the least transmission rate?
(a) a LAN   (b) a MAN   (c) a WAN

1.3 WiMAX is:
(a) a LAN   (b) a MAN   (c) a WAN

1.4 Wi-Fi is:
(a) a LAN   (b) a MAN   (c) a WAN

1.5 The computers in your lab are connected by:
(a) a LAN   (b) a MAN   (c) a WAN

1.6 A cable TV network serving New York City is:
(a) a LAN   (b) a MAN   (c) a WAN

1.7 If you are in Houston, Texas, which network will enable you to share information with your friend in Manchester, United Kingdom?
(a) a LAN   (b) a MAN   (c) a WAN

1.8 Ethernet was invented in the year
(a) 1969   (b) 1973   (c) 1982   (d) 1991

1.9 This is not an organization that develops standard:
(a) ANSI   (b) IEEE   (c) ITU   (d) OSI

1.10 IEEE does not do one of the following:
(a) Develops technical standard   (b) Allocates spectrum   (c) Publishes technical papers   (d) Sponsors conferences

Answer: 1.1 b, 1.2 c, 1.3 b, 1.4 a, 1.5 a, 1.6 b, 1.7 c, 1.8 b, 1.9 d, 1.10 b

Problems

1.1 What is a network protocol?

1.2 What are the differences between a LAN and a WAN?

1.3 Select a LAN and discuss how it works.

1.4 Choose a particular MAN and write a short essay about it.

1.5 The Internet is regarded as a "network of networks." Explain why.

1.6 Write a short essay on either home area network (HAN) or storage area network (SAN).

1.7 The ALOHA network was the first system to successfully use the packet radio for communication. Find out more about ALOHA and write a short essay about it.

1.8 A LAN uses network operating system (NOS) to operate. What specifically does NOS do? Give examples of NOS.

1.9 Why are standards important in computer networking?

1.10 Select a standards organization and write briefly about it.

1.11 What benefits does the government derive in using computer networks?

1.12 Select a social network and write briefly about it.

1.13 Discuss two applications of computer networks not discussed in this chapter.

1.14 Using Web resources, write a short essay on smart cities.

1.15 The Internet Architecture Board published a series of documents known as Requests for Comments (RFCs). Use the Web to find two of the documents.

1.16 Although IBM is not a standards organization, some of its work have become standards. Two examples are Systems Network Architecture (SNA) and the Extended Binary Coded Decimal Interchange Code (EBCDIC). Write about these two standards.

Digital Communications

2

Networking is marketing. Marketing yourself, your uniqueness, what you stand for.

Christine Comaford-Lynch

Abstract

Digital communications is referred to as the transfer of a digital bit stream (data) from point A to point B over a communication channel. It is also known as data transmission. In this chapter, we explore various transmission media, encoding techniques, bit/byte stuffing, multiplexing techniques, and switching mechanisms.

Keywords

Transmission media · Encoding techniques · Bit/byte stuffing · Multiplexing techniques · Switching mechanisms

ily regenerated as compared to analog signals when transmitted from point A to point B. Noise is never a major issue with digital communication but is a major issue in the analog world. The corruption or complete distortion of the signal is always the case in analog communication. The other advantage is the fact that digital circuits are more reliable than analog circuits and are produced at much lower costs. The digital hardware is more flexible in implementation than the analog hardware. This chapter covers various transmission media, encoding techniques, bit/byte stuffing, multiplexing techniques, and switching mechanisms.

Digital data have to be sent from the information source where the sender of the information is located to the information destination where the person to receive the information is located as shown in Fig. 2.1.

Digital communication is the transfer of a digital bit stream (data) from the information source location to the destination location over a communication channel.

2.1 Introduction

In the world where computer networking has become the fastest growing technology in our society, digital communication becomes very important. The question is always asked: why study digital communications? The simple answer lies in the various advantages that digital communication has over analog communication. The primary advantage is the fact that digital signals can be eas-

Information source: **The information source is where the message communicated to the destination originates. It is usually in the form of an information waveform.**

Source encoder: In the source encoder, the information waveforms such as texts, audios, images, and videos are converted to bits.

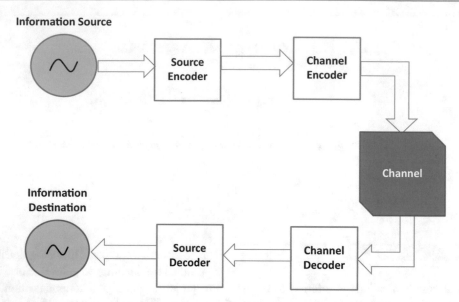

Fig. 2.1 Digital communication system

Source decoder: On the receiving side of the digital communication system, the decoder converts the bits to waveforms.

Channel encoder: The channel encoder is where the bits are converted into signal waveforms.

Channel decoder: The channel decoder converts the received waveform back to bits.

Channel: The information from the source gets to the destination passing through a communication channel. There are different types of communication channels as discussed in Sect. 2.2.

2.2 Transmission Media

The communication channel is referred to as the transmission media. The transmission media provides a physical connection between the transmitter and the receiver.

Transmission media in computer networks is the physical media that allow communication to take place from the source location to the destination location.

There are several examples of these media as shown in Fig. 2.2. They are classified into two categories such as *guided media* and *unguided media*. The guided media consists of twisted pair cable, coaxial cable, and optical fiber cable. The unguided media consists of radio waves, microwaves, and infrared.

2.2.1 Guided Media

The guided transmission media is also known as the wired transmission media or bound transmission media. This type of transmission media consists of cables that are tangible or have physical existence and are limited geographically in a physical way. Twisted pair cable, optical fiber cable, and coaxial cables are some of the popular guided transmission media in use today. Each of them has its own characteristics like physical appearance, transmission speed, cost, and effect of noise.

- *Twisted pair cable:* There are two types of twisted pair cables—unshielded and shielded pair cables.

  - *Unshielded twisted pair (UTP):* It has the capability to block interference. It does this without depending on a physical shield. The unshielded twisted pair is used for

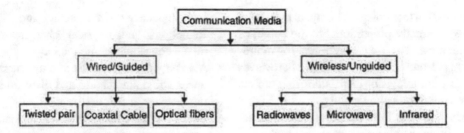

Fig. 2.2 Transmission media type

telephone applications. The advantages of the unshielded twisted pair are that it is easy to install, has high-speed capacity, and costs less. The disadvantages of the UTP are that it only operates more efficiently for short transmissions due to attenuation, can be affected by external interference, and has lower performance and capacity when compared to shielded twisted pair (STP).

- *Shielded twisted pair (STP)*: The STP is a cable that has a special jacket that blocks external interference. It is used in fast data rate Ethernet and in data and voice channels of telephone lines. The advantages are that it is faster compared to other cables, has the ability to eliminate crosstalk, and has better performance at a higher data rate when compared to UTP. The disadvantages are that it is bulky, costs more, and is difficult to install and manufacture when compared to other types of cables.

• *Optical fiber cable*: This type of cable is used for the transmission of large volumes of data. The concept used is that of light through a core made up of plastic or glass. In this case, a less dense plastic or glass covering surrounds the core. The advantages are that it is light in weight, has less signal attenuation, has increased bandwidth and capacity, has electromagnetic interference immunity, and has resistance to corrosive materials. The disadvantages are that it is fragile, is difficulty to install and maintain, operates only in a unidirectional way without additional fiber for bidirectional communication, and is more expensive.

• *Coaxial cable*: This type of cable has an outer plastic that covers two parallel conductors having a separate insulated protection that covers each side. Information is transmitted in two

modes, namely, broadband mode and baseband mode. In the broadband mode, the cable bandwidth is split into separate ranges, while in the baseband mode, the cable bandwidth is dedicated. Coaxial cables are widely used by cable and analog televisions. The advantages of the coaxial cables are that they are easy to install and expand, have better noise immunity, are less expensive, and have high bandwidth. The major disadvantage of the coaxial cable is that once there is a single cable failure the entire network can be disrupted.

2.2.2 Unguided Media

The unguided media is also known as wireless transmission media or unbound transmission media. Data in this case are transmitted without using cables and are not limited geographically in a physical way. In the twenty-first century, wireless communication has become very popular. Wireless LANs are being installed in college campuses as well as offices. Radio wave, microwave, and infrared are some of the popular unbound transmission media.

• *Radio waves*: Radio waves can easily penetrate buildings, and they can be generated easily. The alignment of the sending and receiving antennas is not necessary in this case. The frequency range for radio waves is from 3 KHz to 1 GHz. In their transmission of information from the source location to the destination location, cordless phones and FM and AM radios use radio waves.

• *Microwaves*: In the case of microwaves, the sending and receiving antennas must be properly aligned because it uses the line of sight transmission technology. The signal travel distance is directly proportional to the height of the antenna. The frequency range for microwaves is from 1 GHz to 300 GHz. Microwaves are used for television distribution and mobile phone communication mainly.

- *Infrared waves*: Infrared waves cannot penetrate through obstacles such as buildings. Interference between systems is there prevented. They are only used for very short distance communications. The frequency range for infrared waves is from 300 GHz to 400 THz. This kind of technology is used in printers, TV remotes, keyboards, and wireless mouse.

It should be noted that different media have different properties. Therefore, to choose the best transmission media for any application, the following factors must be considered.

- *Bandwidth:* This refers to the data-carrying capacity of a channel or medium. Lower bandwidth communication channels support lower data rates, while higher bandwidth communication channels support higher data rates.
- *Delay:* Delay consideration of any medium is very important in the transmitting and receiving of information from the source location to the destination location. Transmission delay refers to the time it takes to send the bits of data onto the media and is given by L/R, where L is the amount of data and R is the rate. Propagation delay refers to the time it takes for a bit, once on the media, to reach the destination.
- *Radiation:* This means the consideration of the leakage of the signal from the medium due

lost depends on frequency and distance. Radiations and physical characteristics of media contribute to attenuation.

- *Number of receivers:* Each attachment introduces some attenuation and distortion, limiting distance and/or data rate.
- *Cost:* There are many variables that can influence the cost of implementing a specific type of media. The cost of installation, new infrastructure versus reusing existing infrastructure, lower transmission rate affecting productivity, and cost maintenance and support are some of the key costs to be considered in making any decision.
- *Transmission impairments*: Limit the distance a signal can travel.
- *Interference:* Competing signals in overlapping frequency bands can distort or wipe out a signal.
- *Noise absorption:* It refers to the susceptibility of the media to external electrical noise that can cause distortion of data signal.
- *Ease of installation and maintenance:* The complexity of the installation and maintenance of any transmission media will determine how viable it will be in considering that media for any computer network transmission application.

2.3   Encoding

In digital communications, information that is carried through the transmission media is converted into a format that the communication system can understand and process. The encoding process helps to code the data such that the computer network and the media system can process the information accordingly.

The process of converting data or a given sequence of alphabets, characters, and symbols into a specified format for the secured transmission is called encoding.

to undesirable electrical characteristics of the medium.

- *Attenuation*: It refers to loss of energy as signal propagates outward. The amount of energy

The transmission processing needs that require encoded information include application data processing. For example, the conversion of files, data storage and data compression or decompres-

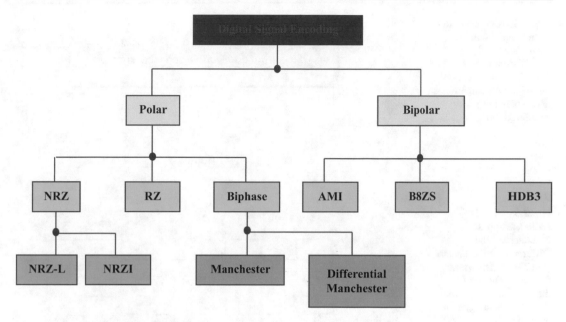

Fig. 2.3 Types of encoding techniques

sion, data transmission, and data program compilation and execution must be done before the information is transmitted to the appropriate communication channel.

Encoding can be categorized in two ways such as:

- Converting analog information to digital information for electronics systems.
- In the case of computer networking technology, the encoding process is applied to the conversion of a specific code, such as numbers, letters, and symbols to equivalent data information.

Data Encoding
The process of encoding data involves using different shapes of current or voltage levels representing *1s* and *0s* of the digital information that has been encoded on the transmission media computer network link.

Encoding Techniques
There are different types of encoding techniques. The most common types are as represented in Fig. 2.3 depending on the type of data conversion. In the encoding of digital signals, two major

encoding techniques, the Polar and the Bipolar, are used. In the Polar encoding, you have Nonreturn to Zero (NRZ), Return to Zero (RZ), and Bi-phase. Within NRZ you have Nonreturn to Zero-Level (NRZ-L) and Nonreturn to Zero-Inverted (NRZ-I). Within Bi-phase, you have the Manchester and the Differential Manchester. As for the Bipolar encoding techniques, you have Alternate Mark Inversion (AMI), Bipolar 8-Zero Substitution (B8ZS), and High-Density Bipolar 3 (HDB3).

Nonreturn to Zero (NRZ)
In the NRZ encoding, it has *0* for low-voltage level and *1* for high-voltage level. The characteristic of the NRZ encoding is that the voltage level remains constant during bit interval. It does not allow the indication of the start or end of a bit while at the same time maintaining the same voltage level as long as the value of the bit previously and currently is the same. Fig. 2.4 indicates how NRZ does its own encoding. As stated earlier, there are two types of NRZ encoding techniques, and they are NRZ-L and NRZ-I.

NRZ-L
This is almost the same as the NRZ, only that there is a change in the polarity of the first bit of

Fig. 2.4 NRZ encoding techniques. (Source: https://www. tutorialspoint.com/ digital_communication/ digital_communication_ data_encoding_ techniques.htm)

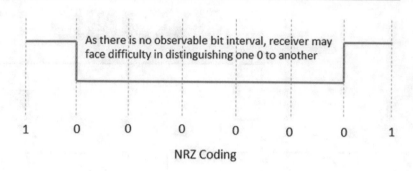

NRZ Coding

Fig. 2.5 NZR-L, NZR-I, Bi-phase Manchester, and Differential Manchester. (Source: https://www. tutorialspoint.com/ digital_communication/ digital_communication_ data_encoding_ techniques.htm)

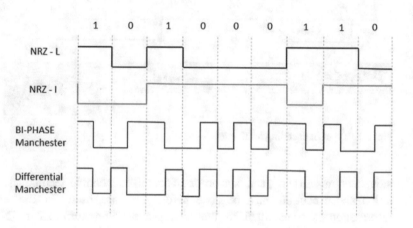

the signal input as shown in Fig. 2.5. When the incoming signal changes from *0* to *1* or from *1* to *0*, the signal changes its polarity.

NRZ-I

In this case, there is no transition at the beginning of the bit interval when there is *0* at the incoming signal. However, there is a transition at the beginning of the bit interval when there is *1* at the incoming signal as shown in Fig. 2.5.

Generally, the results of encoding process of the NRZ have a disadvantage in that the synchronization of the transmitter clock with the receiver clock gets completely disturbed, especially when there is a string of *1s* and *0s*. Therefore, it is advisable to have a separate clock line for this type of encoding process.

Bi-phase Encoding

The Bi-phase encoding process initially checks the signal level twice for every bit time and in the middle as shown in Fig. 2.5. Hence, the clock rate is double the data transfer rate, and thus the mod-

ulation rate is also doubled. The clock is taken from the signal itself. The bandwidth required for this coding is greater.

There are two types of Bi-phase encoding.

- Bi-phase Manchester
- Differential Manchester

Bi-phase Manchester

In the Bi-phase Manchester encoding, the transition is done at the middle of the bit interval as shown in Fig. 2.5. The transition for the resultant pulse is from high to low in the middle of the interval, for the input bit *1*, while the transition is from low to high for the input bit *0*.

Differential Manchester

In the Differential Manchester encoding, the transition takes place in the middle of the bit interval. As shown in Fig. 2.5, if no transition takes place at the beginning of the bit interval, then the input bit is *1*. If transition takes place

Fig. 2.6 Examples of encoding rules for bipolar-AMI, B8ZS, and HDB3. (Source: William, 2014, p. 162)

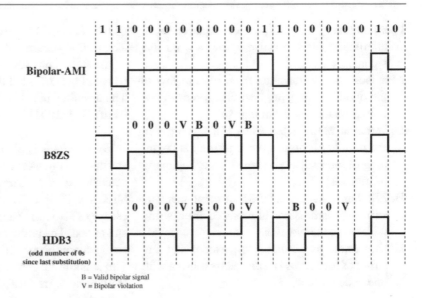

B = Valid bipolar signal
V = Bipolar violation

at the beginning of the bit interval, then the input bit is *0*.

RZ Encoding

The return to zero encoding is done as the signal returns to zero between each pulse. It happens no matter if a number of consecutive *0s* or *1s* occur in the signal. It is a self-clocking signal.

Bipolar Encoding

There are three types of bipolar encoding, namely, Alternate Mark Inversion (AMI), Bipolar 8-Zero Substitution (B8ZS), and High-Density Bipolar 3 (HDB3).

AMI

This type of bipolar encoding system as shown in Fig. 2.6 represents neutral (zero) voltage as binary *0*, and the alternating positive and negative voltages are represented as binary *1*.

B8ZS

The B8ZS has a problem with synchronization being lost when there is a stream of binary *0s* being sent; otherwise, it works very similar as the AMI by changing poles for each binary *1*. It is a common technique used in the USA to avoid the synchronization problem of long strings of binary *0s*. With the artificial signal changes it makes, this problem seems to be fixed. It takes place

when eight consecutive *0s* take place in the bit stream and the signals are known as violations as shown in Fig. 2.6. The violation signal that takes place is based on the polarity of the last binary *1* before the *8 0s* and will match this polarity. Therefore, the receiving end looking for an alternate polarity to the binary *1* will discover the same polarity and will thus determine that there has been a string of *8 0s*. The polarity is set to be the same as the previous positive bit.

HDB3

The HDB3 as shown in Fig. 2.6 uses four zeros rather than eight as in the case of the B8ZS. A violation takes place after the four zeros. Synchronization is made possible and data retrieval more accurate by the use of violations in the signal that give it extra "edges."

2.3.1   Summary of Digital Signal Encoding Formats

Nonreturn to Zero-Level (NRZ-L): 0 = high level, 1 = low level
Nonreturn to Zero-Inverted (NRZ-I): 0 = no transition at beginning of interval (one bit time), 1 = transition at beginning of interval
Bipolar-AMI: 0 = no line signal, 1 = positive or negative level, alternating for successive ones

Pseudoternary: 0 = positive or negative level, alternating for successive zeros, 1 = no line signal

Manchester: 0 = transition from high to low in the middle of interval, 1 = transition from low to high in the middle of interval

Differential Manchester: Always a transition in the middle of interval 0 = transition at beginning of interval, 1 = no transition at beginning of interval

B8ZS: Same as bipolar AMI, except that any string of eight zeros is replaced by a string with two code violations

HDB3: Same as bipolar AMI, except that any string of four zeros is replaced by a string with one code violation

Example 2.1 Figure 2.6 shows the signal encoding for the binary sequence 1100000000110000010 using AMI and then scrambled using B8ZS and HDB3. The original sequence includes a continuous string of eight zeros and five zeros. B8ZS eliminates the string of eight zeros. HDB3 eliminates both strings. The total number of transitions for this sequence is 7 for bipolar-AMI, 12 for B8ZS, and 14 for HDB3.

2.4   Bit/Byte Stuffing

In computer networks, each frame begins and

01101111111001111101111111111100000

The stuffed stream will be as shown below with the stuffed bits (0s) in "red":

01101111**1**011001111100111110111110**0**00000

The unstuffed bits then become:

01101111111001111101111111111100000

Byte stuffing is also referred to as character stuffing in computer networks. The ASCII characters are used as framing delimiters (e.g., DLE STX and DLE ETX) where DLE is data link escape, STX is the start of text, and ETX is the end of text. The problem occurs when these character patterns occur within the "transparent" data. The solution in this case is done by the sender stuffing an extra DLE into the data stream just before each occurrence of an "accidental" DLE in the data stream. The data link layer on the receiving end unstuffs the DLE before giving the data to the network layer.

2.5   Multiplexing

In any digital communication systems or computer network system, information is fed to the network from different sources via a common communication channel. All of this information from different sources is combined before passing through the channel. The device that enables the combination of the various sources is the multiplexer.

Multiplexing is the process in which multiple data streams, coming from different sources, are combined and transmitted over a single data channel or data transmission media.

ends with a special bit pattern called a flag byte [*01111110*] in *bit stuffing*. Whenever the sender data link layer encounters five consecutive ones in the data stream, it automatically stuffs a *0* bit into the outgoing stream. When the receiver sees five consecutive incoming ones followed by a *0* bit, it automatically destuffs the *0* bit before sending the data to the network layer.

Example 2.2 As an example of bit stuffing, if the bit input stream is as follows:

There are two basic forms of multiplexing in digital communication systems or in a computer network system. They are time division multiplexing (TDM) and frequency division multiplexing (FDM). Single channel is divided into non-overlapped time slots in a TDM. Data streams from different sources are divided into units with the same size and interleaved successively into the time slots. In an FDM system, data streams are carried simultaneously on the same transmission medium by allocating to each of them a different

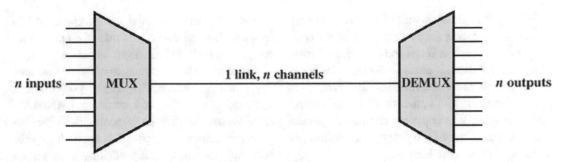

Fig. 2.7   A multiplexing and demultiplexing techniques in a digital communication system

frequency band within the bandwidth of the single channel.

Multiplexing is done by a device called multiplexer (MUX). It is placed at the transmitting end of the communication link. At the receiving end, the composite signal is separated by a device called demultiplexer (DEMUX). Demultiplexer performs the reverse process of multiplexing and routes the separated signals to their corresponding receivers or destinations. The multiplexing and demultiplexing technique in a digital communication system or in a computer network system is shown in Fig. 2.7.

2.6   Switching

There are multiple parts in which information can be transmitted from the source location to the destination location in any computer network. As the information travels through various communication channels, it can be switched to various parts of the communication channel. In digital communication systems, there are four types of switching methods. They are circuit switching, packet switching, message switching, and cell switching.

Circuit Switching
Circuit switching is a method that directly connects the transmitter and the receiver in an unbroken path. As an example, in telephone switching equipment, a path is established that connects the caller's telephone to the receiver's telephone by making a physical connection. Once the path is established, the path is completely dedicated

until the connection is terminated. The circuit switching has the advantage of being a dedicated channel once it is established. The disadvantages are that there is always a long wait before connection is established at least up to 10 secs and could be more for long-distance or international calls. Because of the dedicated channel, it is more expensive than the other types of switching methods.

Packet Switching
Packet switching has two types, namely, virtual circuit and datagram. In the packet switching methods, the information is broken into small parts, called packets. Each packet is tagged with appropriate source and destination addresses. Since packets have a strictly defined maximum length, they can be stored in main memory instead of disk; therefore, access delay and cost are minimized. The transmission speeds, between nodes, are also optimized. With current technology, packets are generally accepted onto the network on a first-come, first-served basis. If the network becomes overloaded, packets are delayed or discarded.

Packet switching has the advantage that it is cost-effective because switching devices do not need massive amount of secondary storage. It also offers improved delay characteristics, because there are no long messages in the queue as the maximum packet size is fixed. If there are busy or disabled links, the packet can be rerouted if there is any problem. In packet switching, many network users can share the same channel at the same time. By making optimal use of link bandwidth, packet switching can maximize link efficiency. The disadvantage of packet switching is that the protocols for packet switching are typically more complex. It can add some initial costs in implementation. If packet is lost, sender needs to retransmit the data. Another disadvantage is that packet-switched systems still cannot deliver the same quality as dedicated circuits in applications requiring very little delay—like voice conversations or moving images.

Message Switching

In the case of message switching, there is no need to establish a dedicated path between two stations quite unlike the circuit switching. The destination address is attached to the message when a message is sent from a station. The message is then transmitted through the network, in its entirety, from node to node. Each node receives the entire message, stores it in its entirety on disk, and then transmits the message to the next node. This type of network is called a store-and-forward network. A message switching node is typically a general-purpose computer. The device needs sufficient secondary storage capacity to store the incoming messages, which could be long. A time delay is introduced using this type of scheme due to store-and-forward time, plus the time required to find the next node in the transmission path.

Some of the key advantages of the message switching are that the channel efficiency can be greater compared to circuit-switched systems because more devices are sharing the channel. The traffic congestion can be reduced, because messages may be temporarily stored in route. The message priorities can be established due to store-and-forward technique, and the message broadcasting can be achieved with the use of broadcast address appended in the message. The disadvantages include the fact that the message switching is not compatible with interactive applications. The store-and-forward devices are expensive because they must have large disks to hold potentially long messages.

Cell Switching

In cell switching, there are similarities with packet switching, except that the switching does not necessarily occur on packet boundaries. This is ideal for an integrated environment and is found within cell-based networks. Cell switching can handle both digital voice and data signals.

Some of the key advantages of the cell switching include scalability, high performance, dynamic bandwidth, and common LAN/WAN architecture multimedia support. High performance is achieved because this technology uses hardware switches. Cell switching uses virtual circuit rather than physical circuit; therefore it is not necessary to reserve network resources for a particular connection. Also, once a virtual circuit is established, switching time is minimized, which ensures higher network throughputs. The cell has a fixed length of 53 bytes out of which 48 bytes are reserved for payloads and 5 bytes act as header. The header contains payload-type information, virtual circuit identifiers, and header error check.

Cell switching has features of circuit switching, as it is a connection-oriented service where each connection during its set-up phase creates a virtual circuit. A connection-oriented service is a technique used to transport data at the session layer. It requires that a session connection be established between the information source location and the information destination location, similar to a phone call. This method is normally considered to be more reliable than a connectionless service, although not all connection-oriented protocols are considered reliable. A connection-oriented service can be a circuit-switched connection or a virtual circuit connection in a packet-switched network. For the latter, traffic flows are identified by a connection identifier, typically a small integer of 10 to 24 bits. This is used instead of listing the destination and source addresses. Cell switching has a major disadvantage, and that is the fact that it is an unreliable, connection-oriented packet-switched data communications protocol.

Summary

1. Digital communications is also known as data transmission and is referred to as the transfer of a digital bit stream (data) from source location to destination location over a communication channel.

2. The encoding process helps to code the data such that the computer network and the media system can process the information accordingly.
3. In the encoding of digital signals, two major encoding techniques, the Polar and the Bipolar are used.
4. In computer networks, each frame begins and ends with a special bit pattern called a flag byte [01111110].
5. Byte stuffing is referred to as character stuffing in computer networks. The ASCII characters are used as framing delimiters (e.g., DLE STX and DLE ETX) where DLE is data link escape, STX is the start of text, and ETX is the end of text.
6. Multiplexing is the process in which multiple data streams, coming from different sources, are combined and transmitted over a single data channel or data stream.
7. In digital communication systems, there are four types of switching methods. They are circuit switching, packet switching, message switching, and cell switching.

2.6 Converting analog information to digital information for electronics systems is one of the ways computer network encoding can be done.
     (a) True   (b) False
2.7 Byte stuffing cannot be referred to as character stuffing in computer networks.
     (a) True   (b) False
2.8 In circuit switching, the long wait before connection is established is at least up to:
     (a) 5 secs   (b) 10 secs   (c) 15 secs   (d) 20 secs
2.9 In the case of message switching, there is no need to establish a dedicated path between two stations quite unlike the circuit switching.
     (a) True   (b) False
2.10 Common LAN/WAN architecture multimedia support is not one of the advantages of cell switching.
     (a) True   (b) False

Answer: 2.1 a, 2.2 b, 2.3 a, 2.4 b, 2.5 d, 2.6 a, 2.7 b, 2.8 b, 2.9 a, 2.10 b

Review Questions

2.1 Digital communications can also be referred to as:
     (a) Data transmission   (b) Analog transmission   (c) Digital bit stream
2.2 The communication channel is also referred to as: the transmission media.
     (a) Communication transmission   (b) Transmission media   (c) Social media
2.3 Computer bus is an example of a transmission media?
     (a) True   (b) False
2.4 Radio wave is not one shape or form of a digital data?
     (a) True   (b) False
2.5 Limiting the distance a signal can travel can be known as:
     (a) Noise absorption   (b) Interference   (c) Attenuation   (d) Transmission impairment

Problems

2.1 Name at least five examples of a transmission media.
2.2 Name five of the different shapes and forms that communication data can be represented.
2.3 What factors must you consider in choosing the best transmission media for any application? Briefly discuss each identified factor.
2.4 Name and discuss the two different types of transmission media.
2.5 What is encoding? What are the two ways encoding can be categorized?
2.6 Describe the Polar and Bipolar encoding techniques.
2.7 Consider the binary sequence 0100101. Draw the waveforms for the following formats.
     (a) Unipolar NRZ signal format
     (b) Bipolar RZ signal format
     (c) AMI (Alternate Mark Inversion) RZ signaling format

2.8 Consider a binary sequence with a long sequence of 1s followed by a single 0 and then a long sequence of 1s such as this binary sequence 11111011111. Draw the waveforms for this sequence, using the following signaling formats:
   (a) Unipolar NRZ signal format
   (b) Bipolar RZ signal format
   (c) AMI (Alternate Mark Inversion) RZ signaling format
   (d) Bi-phase Manchester signaling

2.9 Briefly describe the working of a bit and byte stuffing in a computer network system.

2.10 Given an input bit stream of [01101111111 0011111101111111111100000], determine what the bit stuffing will be.

2.11 Given an input bit stream in a computer network system to be [011011111110011111101111111111100000], determine what the bit stuffing will be.

2.12 What will be the unstuffed bit stream at the output end of the computer network system?

2.13 (a) Define the multiplexing process. (b) Briefly describe the two forms of a multiplexer in a digital communication system.

2.14 (a) How does circuit switching work in a computer network system? (b) What are the advantages and disadvantages of a circuit switching?

2.15 What are the (a) advantages and (b) disadvantages of a packet switching?

2.16 What is the difference between message switching and circuit switching?

2.17 (a) What are the differences between cell switching and packet switching? (b) What are the advantages and disadvantages of cell switching?

Network Models

3

Networking is an investment in your business. It takes time and when done correctly can yield great results for years to come.

Diane Helbig

Abstract

The International Standards Organization (ISO) created the Open Systems Interconnection (OSI) model, which covers all aspects of network communication standards. The entire data communication relies on two computer network models called the Open Systems Interconnection (OSI) and the Transmission Control Protocol/Internet Protocol (TCP/IP) models. We cover in this chapter the OSI model, the TCP/IP models, and all the different layers associated with the models. We cover also the comparisons between the OSI and the IEEE Layer models, the functionalities of their different layers, and how they relate to each other.

Keywords

Open Systems Interconnection (OSI) · Transmission Control Protocol/Internet Protocol (TCP/IP) model · IEEE Layer models · Internet Control Message Protocol (ICMP) · Signaling System No. 7 (SS7)

3.1 Introduction

The computer network models have the function of sending information from a sender to a receiver and making sure the desired information reaches its destination without any disruption. The entire data communication relies on two computer network models called the Open Systems Interconnection (OSI) and the Transmission Control Protocol/Internet Protocol (TCP/IP) models.

In this chapter, we will discuss the OSI model that was developed in 1983 by the International Standards Organization (ISO). The OSI model allows you to know the different functionalities in network communications. We will also cover the TCP/IP models and all the different layers associated with the model. The TCP/IP model is slightly different from the OSI model, and those differences will be discussed. This chapter will discuss different networking applications—the internetworking system devices and Signaling Systems No. 7 (SS7). Their various application areas will be explored. Finally, we will focus on the comparisons between OSI and the IEEE Layer models, the functionalities of their different layers, and how they relate to each other.

3.2   OSI Model

The International Standards Organization (ISO) which is a worldwide multinational body was established in 1947. Its major task was to work on international standards agreements. The ISO created the Open Systems Interconnection (OSI) model which covers all aspects of network communication standards. The OSI model first came into existence in the late 1970s. The OSI model consists of seven layers as shown in Fig. 3.1. These layers are:

1. The physical layer
2. The data link layer
3. The network layer
4. The transport layer
5. The session layer
6. The presentation layer
7. The application layer

Layers 1–4 relate to communications technology, while layers 5–7 relate to user applications. In the OSI model, each layer provides services to layer above and "consumes" services provided by layer below. Active elements in a layer are

called entities. Entities in the same layer in different machines are called peer entities. In OSI models, layers can offer connection-oriented or connectionless services. Connection-oriented services are like telephone system. Connectionless services are like postal system.

Each service has an associated quality of service (e.g., reliable or unreliable). Quality of service (QoS) refers to any technology that manages data traffic to reduce latency, jitter on the network, and packet loss. There has to be some level of predictability in the network, and QoS assures that in any network. Enterprise networks need to provide predictable and measurable services as applications such as voice, video, and delay-sensitive data traverse the network. QoS is also the description or measurement of the overall performance of a service, such as a telephony, computer network, and cloud.

3.2.1   The Physical Layer

The function of the physical layer is to transmit individual bits from one computer to another. It also makes sure as a matter of concern with the transmission of unstructured bit stream over the physical medium that the electrical, functional, mechanical, and procedural characteristics access the physical medium. The physical layer regulates the transmission of a stream of bits over a physical medium. It defines how the cable is attached to the network adapter and what transmission technique is used to send data over the cable. It deals with issues like:

- The definition of 0 and 1, for example, how many volts represent a 1 and how long a bit lasts
- Whether the channel is simplex or duplex
- How many pins a connector has and what the function of each pin is

3.2.2   The Data Link Layer

The data link layer is responsible for transmitting of frames from one computer to another. It pro-

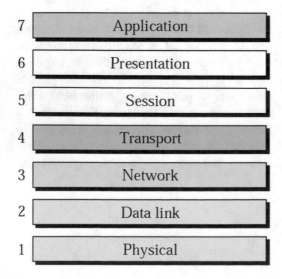

Fig. 3.1 The OSI model

7	Application
6	Presentation
5	Session
4	Transport
3	Network
2	Data link
1	Physical

vides reliable transfer of information across the physical link. It makes sure that the frames it sends have the necessary synchronization, flow control, and error control. This layer takes the "raw" transmission facility provided by the physical layer (Layer 1) and uses it to provide a reliable, error-free transmission service. It does this by breaking the data stream up into frames, typically of thousands of bytes in length. Where necessary, special acknowledgment frames can be sent back by the receiving data link entity to indicate successful receipt of each frame.

The physical layer (Layer 1) transmits a continuous sequence of bits, and so Layer 2 must create and recognize frame boundaries. This is typically done by the insertion of special bit patterns. Channel errors can completely destroy a frame, and hence retransmission may be necessary. This, in turn, leads to the possibility of duplicate frames being received. Data link protocols deal with issues such as channel errors using the idea of duplication. A data link protocol may offer several different "service classes" to the network layer, each with a different quality and cost. The data link layer must also regulate traffic flow to prevent "swamping" of a slow receiver. In a duplex channel, there may be competition between data and acknowledgment frames; one solution is known as piggybacking, where acknowledgment information is attached to data frames.

3.2.3   The Network Layer

The network layer is responsible for the delivery of packets from the original source to the final destination. This layer is also responsible for providing upper layers with independence from the data transmission and switching technologies used to connect to the systems. It is responsible for establishing, maintaining, and terminating connections. The network layer controls the subnet. A subnet, also called a subnetwork, is a logical subdivision of an IP network. The practice of dividing a network into two or more networks is called subnetting. Computers that belong to a subnet are addressed with an identical most significant bit-group in their IP addresses. The key

design issue is routing of data within the subnet. Routing can be based on static tables (rarely changed), determined at the start of each session, or highly dynamic: individually determined for each packet, reflecting the current network load.

Too many packets in the subnet simultaneously can form bottlenecks; this is dealt with by congestion control procedures in the network layer. Accounting for subnet use is also typically the responsibility of the network layer. When a packet travels from one network to another, address conversion may be necessary. The address conversion is performed by the network layer. In broadcast networks (e.g., many LANs), the routing problem is very straightforward, and hence Layer 3 will be very simple, or even nonexistent.

3.2.4   The Transport Layer

The transport layer is responsible for the delivery of a message from one process to another. It provides a reliable, transparent transfer of data between end points. It also provides end-to-end flow control and error recovery. The basic function of the transport layer is to take data from the session layer (Layer 5), split it up into smaller units if necessary, and pass these units on to the network layer. It is then also responsible for ensuring that all the pieces are received correctly and reassembled in the correct order.

Typically the transport layer will create a distinct network layer connection for each transport connection requested by the session layer. However, depending on the data load and the capacity of a single session channel: multiple network connections might be used to support a single high-bandwidth session connection, or one high-bandwidth network connection might be used to support several session connections. The transport layer also determines what type of service to provide to the session layer and, ultimately, to the network users, for example, an error-free, point-to-point channel, guaranteeing data is delivered in the correct order (the most common type of service), transport of isolated messages with no guarantee of correct ordering, or message broadcast to multiple destinations.

Transport is the first true end-to-end layer, i.e., the transport protocol communicates between end parties and not to any of the intermediaries.

3.2.5   Session Layer

The session layer provides the control structure for communication between applications. It establishes, manages, and terminates connections (sessions) between cooperating applications. It makes it possible for applications on different computers to establish, use, and end a session. An example is in the provision of a control structure in the file transfer and remote login on different computer applications. It establishes dialog control and regulates which side transmits, plus when and how long it transmits. It performs token management and synchronization.

3.2.6   Presentation Layer

The presentation layer provides independence to the application processes from differences in data representation (syntax). The presentation layer is concerned with the representation of data to be transmitted. For example, it provides a standard encoding for data. Different computers often use different representations for data structures such as characters (ASCII or EBCDIC), integers (one's complement or two's complement and varying byte ordering conventions), and floating point values (IEEE or proprietary). The presentation layer provides a standard encoding technique (using ASN.1—Abstract Syntax Notation 1). The presentation layer is also concerned with other representation issues such as data compression and encryption of data. In terms of the usual communications model, the presentation layer is responsible for source coding of data.

3.2.7   Application Layer

The application layer provides access to the OSI environment for users and also provides distributed information services. It is used for a wide variety of protocols to meet specific user needs. As an example, network virtual terminal used by "networked screen editors" and other networked applications need to handle many different terminal types. In application layer, software exists to map the functions of the virtual terminal onto the real terminal. Examples include OSI virtual local area networks (VLAN), Trunking Protocol (VTP), and Internet telnet. Another example is file transfer where different file systems have different file naming conventions and internal record structures. Such a protocol must convert files to a "standard" representation. Examples include OSI file transfer, access and management (FTAM), and Internet ftp. In electronic mail, examples of electronic mail protocols are OSI X.400 and Internet mail. In directory service, examples include OSI X.500.

3.3   TCP/IP Model

TCP/IP means Transmission Control Protocol and Internet Protocol. It was developed by the Department of Defense's Project Research Agency (ARPA, later called DARPA) as a part of a research project of network interconnection to connect remote machines. The features that were very prominent during the research, which led to making of the TCP/IP reference model, were:

- The idea of having support for a flexible architecture. Adding more machines to a network was easy.
- The idea of having a robust network and connections that should remain intact until the source and destination machines were functioning.

The overall idea was to allow one application on one computer to talk to (send data packets) to another application running on different computer. The TCP/IP model has five layers as shown in Fig. 3.2. These layers are (1) physical layer, (2) network layer, (3) Internet layer, (4) transport layer, and (5) application layer. The TCP/IP model describes a set of general design

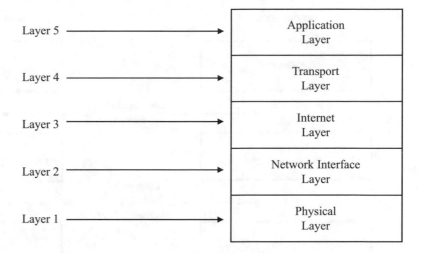

Fig. 3.2 The five layers of the TCP/IP reference model

Layer 5 ———————→ Application Layer

Layer 4 ———————→ Transport Layer

Layer 3 ———————→ Internet Layer

Layer 2 ———————→ Network Interface Layer

Layer 1 ———————→ Physical Layer

guidelines and implementations of specific networking protocols to enable computers to communicate over a network. TCP/IP provides end-to-end connectivity specifying how data should be formatted, addressed, transmitted, routed, and received at the destination. Figure 3.3 is the TCP/IP architectural model showing the connectivity between the TCP/IP source and destination hosts.

3.3.1   Physical Layer

The TCP/IP physical layer is similar to the OSI physical layer. It looks out for hardware addressing, and the protocols present in this layer allow for the physical transmission of data. It defines details of how data is physically sent through the network, including how bits are electrically or optically signaled by hardware devices that interface directly with the network medium, such as coaxial cable, optical fiber, or twisted-pair copper wire. The protocols included in physical layer are Ethernet, token ring, FDDI, X.25, frame relay, etc. The most popular LAN architecture among those listed above is Ethernet. Ethernet uses an access method called CSMA/CD (Carrier Sense Multiple Access/Collision Detection) to access the media, when Ethernet operates in a shared media. An access method determines how a host will place data on the medium.

3.3.2   Network Interface Layer

The network interface layer is responsible for the specification of how to organize data into frames over a network. It is concerned with getting packets from source to destination. The network interface layer must know the topology of the subnet and choose appropriate paths through it. When the source and the destination are in different networks, the network layer (IP) must deal with these differences. The network interface layer controls the network hardware. It performs mapping from IP addresses to hardware addresses. It encapsulates and transmits outgoing packets. It accepts and demultiplexes incoming packets. The software in the network interface layer provides a network interface abstraction, which defines the interface between the protocol software in the operating system and the underlying hardware. It hides hardware details and allows protocol software to interface with a variety of network hardware using the same data structures.

3.3.3   Internet Layer

The Internet layer is where the IP protocol is active and adds the source and destination IP addresses to the packets and routes them to the recipient computer. It specifies the format of the packets sent across an Internet as well as the

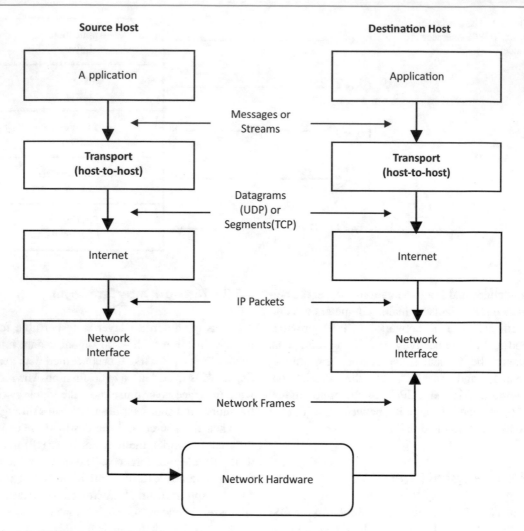

Fig. 3.3 TCP/IP architectural model

mechanisms used to forward the packets from a computer through one or more routers to a final destination. The IP protocol that is used in this layer is the most significant part of the entire TCP/IP suite. The other protocols are the Address Resolution Protocol (ARP) and the Internet Control Message Protocol (ICMP). IPv6 is another new protocol of the Internet layer replacing IPv4 protocol.

The responsibilities of the IP protocol are as follows:

- IP addressing: With this protocol, the logical host addresses known as IP addresses are implemented. The IP addresses are used by the Internet and higher layers to identify the device and to provide internetwork routing.

- Host-to-host communication: It determines the path through which the data is to be transmitted.

- Data encapsulation and formatting: An IP protocol accepts the data from the transport layer protocol. It ensures that the data is sent and received securely. It encapsulates the data into message known as IP datagram.

- Fragmentation and reassembly: The limit imposed on the size of the IP datagram by data link layer protocol is known as maximum transmission unit (MTU). If the size of IP datagram is greater than the MTU unit, then the IP protocol splits the datagram into smaller units so that they can travel over the local network. Fragmentation can be done by the sender or intermediate router. At the receiver side, all the fragments are reassembled to form an original message.

- Routing: When IP datagram is sent over the same local network such as local area network (LAN), metropolitan area network (MAN), and wide area network (WAN), it is known as direct delivery. When source and destination are on the distant network, then the IP datagram is sent indirectly. This can be accomplished by routing the IP datagram through various devices such as routers.

The responsibilities of the Address Resolution Protocol (ARP) are as follows:

- It is the network layer protocol which is used to find the physical address from the IP address. The two terms that are associated with the ARP Protocol are:
  - ARP request: When a sender wants to know the physical address of the device, it broadcasts the ARP request to the network.
  - ARP reply: Every device attached to the network will accept the ARP request and process the request, but only the recipient recognizes the IP address and sends back its physical address in the form of ARP reply. The recipient adds the physical address both to its cache memory and to the datagram header.

The responsibilities of the Internet Control Message Protocol (ICMP) are as follows:

- It is a mechanism used by the hosts or routers to send notifications regarding datagram problems back to the sender.
- A datagram travels from router-to-router until it reaches its destination. If a router is unable to route the data because of some unusual conditions such as disabled links, a device is on fire, or network congestion, then the ICMP protocol is used to inform the sender that the datagram is undeliverable.

An ICMP protocol mainly uses two terms:

- ICMP test: This is used to test whether the destination is reachable or not.
- ICMP reply: It is used to check whether the destination device is responding or not.

It should be noted that the core responsibility of the ICMP protocol is to report the problems, not correct them. The responsibility of the correction lies with the sender. ICMP can send the messages only to the source, but not to the intermediate routers because the IP datagram carries the addresses of the source and destination but not of the address of the router that it is passed to.

IPv6 Protocol: A New Internet Protocol

The IPv6 is the new Internet protocol that replaced the old IPv4 protocol. It has been a backbone of the networks generally and Internet at large. IPv6 is a total replacement in making more IP addresses available. IPv6 addresses are 128 bits in length. The addresses are assigned to individual interfaces on nodes, not to the nodes themselves. A single interface may have multiple unique unicast addresses. Any of the unicast addresses associated with a node's interface may be used to uniquely identify that node. There are some fundamental changes to the protocol that must be considered in security policy. Changing to IPv6 reduces the risks of cybercrime within the network.

3.3.4 Transport Layer

The transport layer is where the Transmission Control Protocol (TCP) is active and is concerned with host-to-host communications. It is responsible for the reliability, flow control, and correction of data which is being sent over the network. The language settings and packed size are agreed. The packets are also checked to confirm safe delivery through acknowledgment. User Datagram Protocol (UDP) is an alternative to TCP where packets are just sent without acknowledgment. The two protocols used in the transport layer are TCP and UDP.

Transmission Control Protocol (TCP)

It provides full transport layer services to applications. It creates a virtual circuit (VC) between the sender and receiver, and it is active for the duration of the transmission. A virtual circuit is a means of transporting data over a packet-switched computer network in such a way that it appears as though there is a dedicated physical layer link between the source and destination end systems of this data. The term VC is synonymous with virtual connection and virtual channel. TCP is a reliable protocol as it detects the error and retransmits the damaged frames. It therefore ensures all the segments must be received and acknowledged before the transmission is considered complete and virtual circuit discarded. At the sending end, TCP divides the whole message into smaller units known as segments, and each segment contains a sequence number which is required for reordering the frames to form an original message. At the receiving end, the TCP collects all the segments and reorders them based on sequence numbers.

User Datagram Protocol (UDP)

The UDP provides connectionless service and end-to-end delivery of transmission. It is an unreliable protocol as it discovers the errors but not specify the error. The UDP discovers the error, and the ICMP protocol reports the error to the sender that user datagram has been damaged.

UDP consists of the following fields as shown in Fig. 3.4:

- Source port address: It is the address of the application program that has created the message.
- Destination port address: It is the address of the application program that receives the message.
- Total length: It defines the total number of bytes of the user datagram in bytes.
- Checksum: It is a 16-bit field used in error detection.

UDP does not specify which packet is lost. It contains only checksum which is the 16-bit field used for error checking of the header and data. It does not contain any ID of a data segment.

3.3.5   Application Layer

In the application layer, all applications such as web browsers and email clients take place. The requests are also made in the application layer to web servers, and emails are originated. Once these requests are made, they are then passed on to the transport layer. The application layer is the topmost layer in the TCP/IP model. It is responsible for handling high-level protocols and issues of representation. The user is allowed to interact with the application in this layer. When one application layer protocol wants to communicate with another application layer, it forwards its data to the transport layer. It is important to note that not every application can be placed inside the application layer with the exception of those who interact with the communication system. As an

Fig. 3.4 An illustration of the UDP with four fields. (Source: https://www.ipv6.com/general/udp-user-datagram-protocol/)

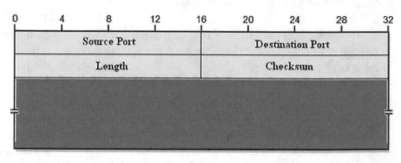

example, the text editor cannot be considered in the application layer while the web browser is using the HTTP protocol to interact with the network. The HTTP protocol ia an application layer proptocol. Some of the protocol examples in the application layer are:

- Hypertext transfer protocol (HTTP). This protocol allows the user to access the data over the World Wide Web (www). It transfers the data in the form of plain text, audio, and video. It is known as a hypertext transfer protocol as it has the efficiency to use in a hypertext environment where there are rapid jumps from one document to another.

3.4 Application I: Internetworking Devices

Networks in any system grow over time sometimes making it very difficult for better performance. Too many network hosts can cause network traffic congestion, and when too many packets are transmitted on a network, the performance of the network degrades to an extent that no packet is delivered. To avoid network traffic congestion, a large network is usually broken into small network segments. The breaking of a network into smaller network segments is called subnetting, and it makes it easier for internetworking using appropriate devices.

An internetworking device is a widely-used term for any hardware within networks that connect different network resources.

- Simple network management protocol (SNMP). This is the framework used for managing the devices on the Internet by using the TCP/IP protocol suite.
- Simple mail transfer protocol (SMTP). The TCP/IP protocol that supports the email is known as a simple mail transfer protocol. This protocol is used to send the data to another email address.
- Domain name system (DNS). An IP address is used to identify the connection of a host to the Internet uniquely. However, users prefer to use the names instead of addresses. Therefore, the system that maps the name to the address is known as domain name system.
- Terminal network (TELNET). It establishes the connection between the local computer and remote computer in such a way that the local terminal appears to be a terminal at the remote system.
- File transfer protocol (FTP). It is a standard Internet protocol used for transmitting the

All devices have separately installed scope features, per network requirements and scenarios. Networking and internetworking devices are divided into seven types: hubs, bridges, switches, routers, gateways, brouters, and repeaters.

3.4.1 Repeaters

Repeaters as shown in Fig. 3.5 operate at the physical layer. Their job is to regenerate the signal over the same network before the signal becomes too weak or corrupted so as to extend the length to which the signal can be transmitted over the same network. The important point to be noted about repeaters is that they do not amplify the signal. When the signal becomes weak, they copy the signal bit by bit and regenerate it at the original strength. It is a two-port device.

3.4.2 Hubs

A hub is an electronic device that implements a network.

files from one computer to another computer.

When computers are connected to a hub as shown in Fig. 3.6, they can communicate as if they are

Fig. 3.5 Illustration of
a repeater system

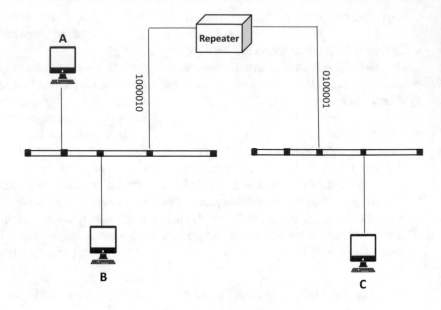

Fig. 3.6 Illustration of
a computer hub

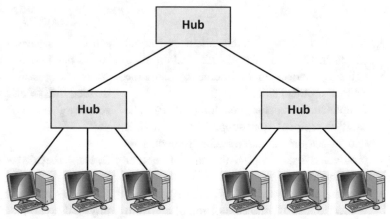

attached to a network. A hub is a multiport repeater. It is normally used to create connections between stations in a physical start topology. A network hub has no routing tables or intelligence on where to send information, and all network data are broadcasted across each connection. There are two types of hubs—the active hub and the passive hub.

- *Active hub*: The active hubs have their own power supply and can clean, boost, and relay the signal along with the network. These active hubs can serve as both as a repeater and as a wiring center. They are also used to extend the maximum distance between nodes.

- *Passive hub*: The passive hubs collect wiring from nodes and power supply from active hub. These passive hubs relay signals onto the network without cleaning and boosting them and cannot be used to extend the distance between nodes like the active hubs.

3.4.3 Bridges

A bridge as shown in Fig. 3.7 operates at the data link layer (Layer 2). A bridge is a repeater, which adds on the functionality of filtering content by reading the MAC addresses of source and destination. It is also used for interconnecting two LANs working on the same protocol. It has a

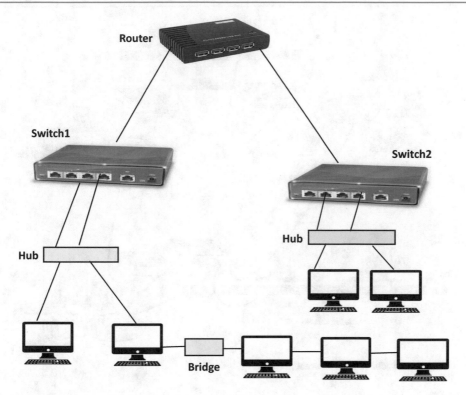

Fig. 3.7 An illustration of repeaters, hubs, switches, and bridges

single input and single output port, thus making it a two-port device. The following types of bridges are common.

- *Transparent bridges*: These are the bridges in which the stations are completely unaware of the bridge's existence, i.e., whether or not a bridge is added or deleted from the network, reconfiguration of the stations is unnecessary. These bridges make use of two processes, i.e., bridge forwarding and bridge learning.
- *Source routing bridges*: In these bridges, routing operation is performed by source station, and the frame specifies which route to follow. The host can discover frame by sending a special frame called discovery frame, which spreads through the entire network using all possible paths to destination.
- *Translational bridge*: The translational bridge converts the data format of one networking to another, for example, converting token ring to Ethernet and vice versa.

3.4.4   Switches

Switches as shown in Fig. 3.7 are multiport bridges with buffers and designs that can boost their efficiencies (large numbers of ports imply less traffic) and performances. Switches are data link layer devices. The switches can perform error checking before forwarding data that make them very efficient as they do not forward packets that have errors and forward good packets selectively to correct ports only. That means that switches divide collision domains of hosts, but broadcast domains remain the same.

3.4.5   Routers

Routers as shown in Figs. 3.7 and 3.8 are devices like switches that route data packets based on their IP addresses. Routers are mainly network layer devices. Routers normally connect LANs and WANs together and have a dynamically updating routing tables based on which they

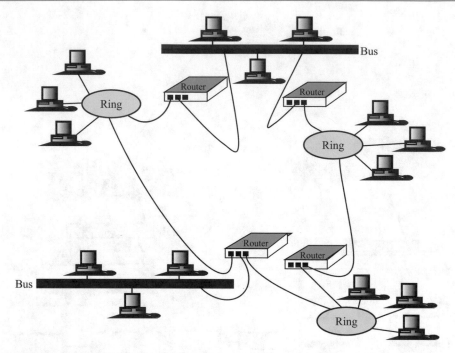

Fig. 3.8 Illustration of routers in an Internet network

make decisions on routing the data packets. Routers divide broadcast domains of hosts connected through them. The two ways through which routers can receive information are as follows:

- *Static routing*: In static routing, the routing information is fed into the routing tables manually. It does not only become a time-consuming task but gets prone to errors as well. The manual updating is also required in case of statically configured routers when change in the topology of the network or in the layout takes place. Thus static routing is feasible for the tiniest environments with minimum of one or two routers.
- *Dynamic routing*: In the case of larger environments, the dynamic routing proves to be the practical solution. The process involves the use of peculiar routing protocols to hold communication. The purpose of these protocols is to enable the other routers to transfer information about the other routers, so that the other routers can build their own routing tables.

3.4.6 Gateways

Gateways as shown in Fig. 3.9 are passages that are used to connect two networks together that may work upon different networking models. They basically work as the messenger agents that take data from one system, interpret them, and transfer them to another system. Gateways are also called protocol converters and can operate at any network layer. Gateways are generally more complex than switches or routers.

3.4.7 Brouters

Brouters are also known as bridging routers and are devices which combine features of both bridges and routers as shown in Fig. 3.10. They can work either at the data link layer or at network layer. Working as routers, they are capable of routing packets across networks, and working as bridges, they are capable of filtering local area network traffics.

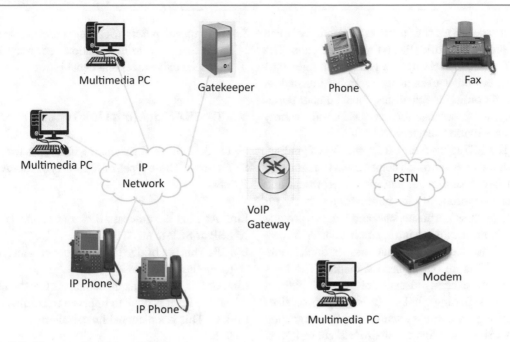

Fig. 3.9 Diagram of a gateway system

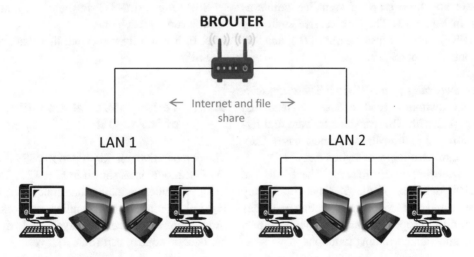

Fig. 3.10 Illustration of a brouter

3.5   Application II: Signaling Systems No. 7 (SS7)

The Signaling System No. 7 (SS7) is a world-wide standard for telecommunications defined by the International Telecommunication Union (ITU) Telecommunication Standardization Sector (ITU-T). It was developed in 1975. The SS7 standard defines the procedures and protocol by which network elements in the public switched telephone network (PSTN) exchange information over a digital signaling network to enable wireless (cellular) and wireline call setup, routing, and control. It is data communications network standard intended to be used as a control and management network for telecommunication networks. It provides call management, database query, routing, and flow and congestion control functionality for telecommunication networks. It is a set of telephony signaling protocols that are used to set up most of the world's public switched telephone network (PSTN) telephone calls. The SS7 primarily sets up and tears down telephone calls, but other uses include number translation, prepaid billing mechanisms, local number portability, short message service (SMS), and a variety of mass market services.

The SS7 takes a digital approach to signaling. With the new approach comes great functionality and better service. It can meet the present and future requirements for information transfer for inter-processor transactions, and it has to provide a reliable means of information transfer in correct sequence and without loss or duplication. The SS7 is an out-of-band signaling. The messages are exchanged over 56K or 64K bidirectional signaling links. It supports services requiring database systems. The SS7 signaling systems allow for an improved control over fraudulent network usage.

There are three types of signaling points as shown in Fig. 3.11. They are service switching point (SSP), signal transfer point (STP), and service control point (SCP).

Service switching points (SSPs): These are telephone switches (end offices or tandems) equipped with SS7-capable software and terminating signaling links. In most cases, they originate, terminate, or switch calls.

Signal transfer points (STPs): These are the packet switches of the SS7 network. They receive and route incoming signaling messages toward the proper destination. They also perform specialized routing functions.

Service control points (SCPs): These are databases that provide information necessary for advanced call-processing capabilities.

3.5.1 SS7 Signaling Link Types

Figure 3.12 shows the signaling link types for the SS7 system. These types of links range from A to F links.

Link A: This is an access link that connects an SSP or SCP to an STP.

Link B: This is a bridge link that connects an STP to another STP.

Link C: It is a cross link that connects two "mated pair" STP. It is done to improve reliability.

Link D: This is a diagonal link similarly to the B link.

Link E: It is an extended link that connects an SSP to an extra STP in the event that the A link cannot reach one.

Link F: It is a fully associated link that connects SSPs.

3.5.2 Comparative Relationship of SS7 to OSI

The comparative relationship of the SS7 model to the OSI model is as shown in Fig. 3.13. The SS7 model includes signaling data link layer, signaling link control layer, signaling network functions, connection control layer, and signaling connection control part (SCCP) layer.

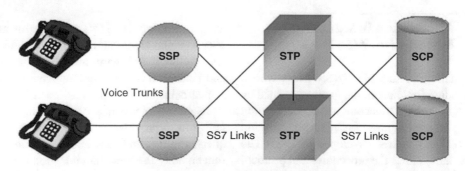

Fig. 3.11 SS7 signaling point diagram

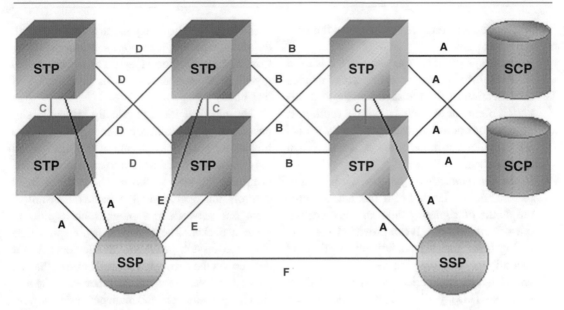

Fig. 3.12 SS7 signaling links

Fig. 3.13 Diagram showing the relationship between the SS7 and OSI models

OSI Model

7	**Application**
6	**Presentation**
5	**Session**
4	**Transport**
3	**Network**
2	**Data Link**
1	**Physical**

SS7

SCCP (Signaling Connection Control)
Signaling Network Functions
Signaling Link Control
Signaling Data Link

- *SS7 data link layer: layer 1*—This is the same as the physical layer in the OSI model. It has two bidirectional transmission paths for signaling. It also has two data channels running in opposite directions at the same data rate which is usually 56K to 64K.
- *SS7 signal link layer: layer 2*—This layer is responsible for signal unit alignment and delimitation. It does error detection and correction using a preventive cyclic retransmission method. It is also responsible for flow control within the system. There are three types of SS7 signaling units. They are (1) the

message signal unit (MSU) that carries the signaling information, (2) the link station signal unit (LSSU) that carries the link status information, and (3) the fill-in signal unit (FISU), which fills in the idle time.
- *SS7 network functions: layer 3*—This layer does the *message handling functions* such as *message routing function* that is used at each signaling point to determine the outgoing link. It does the *message discrimination function* which determines if a received message is destined for a particular point. It also does the *message distribution function*

which delivers received message to the correct user part.

- In the *traffic management functions*, it is used to control the message routing. This includes modification of the message routing to destination points or to continue normal routing. It is used for control signaling traffic by avoiding irregularities in the message. It is also used flow control.
- In the *link management functions*, it is used to restore the availability of a link set, monitor the status of signaling link, and also correct out-of-service links. It can provision new signaling links, and it automatically configures/reconfigures signaling links.
- *SS7 signaling connection control part (SCCP): layer 4*—This layer is responsible for services that are connected oriented, connectionless, and peer-to-peer communication which includes the establishment of connection, the starting of data transfer, and the releases of the connections.

3.6   OSI Model and the IEEE Model

Figure 3.14 shows the OSI model and the IEEE model. In 1983, the International Standards Organization (ISO) developed a network model called Open Systems Interconnection (OSI) reference model, which defined a framework of computer communications. The OSI model has

seven layers. These layers are the physical, data link, network, transport, session, presentation, and application layers. These OSI model layers have been described in Section 3.2 of this book.

The physical layer physically transmits signals across a communication medium. The data link layer transforms a stream of raw bits (0s and 1s) from the physical layer into an error-free data frame for the network layer. The network layer controls the operation of a packet transmitted from one network to another, such as how to route a packet. The transport layer splits data from the session layer into smaller packets for delivery on the network layer and ensures that the packets arrive correctly at the other end. The session layer establishes and manages sessions, conversions, or dialogs between two computers. The presentation layer manages the syntax and semantics of the information transmitted between two computers. The application layer, the highest layer, contains a variety of commonly used protocols, such as file transfer, virtual terminal, and email. The ISO/OSI model requires that the function of each layer define the international standardized network protocols.

In the IEEE LAN model, there are four layers. They are the physical, medium access, logical link, and the higher-level protocols. The physical layer is responsible for transmitting and receiving bits across the various types of media. The performance functions needed to control access to the physical medium are done by the media

Fig. 3.14 Diagram showing the relationship between the OSI model and the IEEE model

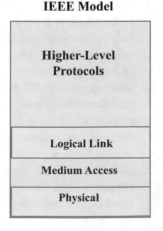

access control layer. The layer responsible for the common interface to the higher software layers is the logical link control layer. Both the OSI and the IEEE are governed by the IEEE 802 reference models. The Institute of Electrical and Electronics Engineers (IEEE) developed a set of LAN standards, known as IEEE Project 802, which the ISO accepted as international standards. The IEEE LAN standards addressed only the lowest two layers, the physical and data link layers, of the ISO/OSI model.

Summary

1. The computer network models have the function of sending information from a sender to a receiver and making sure the desired information reaches its destination without any disruption.
2. The OSI model consists of seven layers which are physical layer, data link layer, network layer, transport layer, session layer, presentation layer, and application layer.
3. The physical layer transmits individual bits from one computer to another.
4. The data link layer transmits frames from one computer to another.
5. The network layer is responsible for the delivery of packets from the original source to the final destination.
6. The transport layer is responsible for the delivery of a message from one process to another.
7. The session layer provides the control structure for communication between applications.
8. The presentation layer provides independence to the application processes from differences in data representation (syntax). It is concerned with the representation of data to be transmitted, for example, providing a standard encoding for data.
9. The application layer provides access to the OSI environment for users and also provides distributed information services.
10. The TCP/IP model has five layers which are physical layer, network layer, Internet layer, transport layer, and application layer.

11. To avoid network traffic congestion, a large network is usually broken into small network segments. The breaking of a network into smaller network segments is called subnetting, and it makes it easier for internetworking using appropriate devices.
12. Networking and internetworking devices are divided into seven categories: hubs, bridges, switches, routers, gateways, brouters, and repeaters.
13. The Signaling System No. 7 (SS7) is a data communications network standard intended to be used as a control and management network for telecommunication networks.
14. There are three types of signaling points which are service switching point (SSP), signal transfer point (STP), and service control point (SCP).
15. The SS7 model includes signaling data link layer, signaling link control layer, signaling network functions, connection control layer, and signaling connection control part (SCCP) layer.
16. In the IEEE LAN model, there are four layers. They are the physical, medium access, logical link, and the higher-level protocols.

Review Questions

3.1 How many layers does the OSI model have?
   (a) 2
   (b) 7
   (c) 5
   (d) Less than 7
3.2 In the OSI model, layers 1–4 relate to communications technology.
   (a) True
   (b) False
3.3 In the OSI model, layers 5–7 do not relate to user applications.
   (a) True
   (b) False
3.4 In the TCP/IP model, the network interface layer is responsible for the specification of how to organize data into frames over a network.

(a) True

(b) False

3.5 The responsibilities of the ARP protocol are:

(a) It is the network layer protocol which is used to find the Internet address from the IP address.

(b) It is the network layer protocol which is used to find the application address from the IP address.

(c) It is the network layer protocol which is used to find the session address from the IP address.

(d) It is the network layer protocol which is used to find the network address from the IP address.

(e) It is the network layer protocol which is used to find the physical address from the IP address.

3.6 Hubs can be classified as:

(a) Non-active

(b) Passive

(c) Active and passive

(d) Non-passive

3.7 Gateways are also called protocol converters and can operate at any network layer.

(a) True

(b) False

3.8 The Signaling System No. 7 (SS7) is a data communications network standard not intended to be used as a control and management network for telecommunication networks.

(a) True

(b) False

3.9 The types of signaling points are:

(a) Service switching point (SSP) and signal transfer point (STP)

(b) Signal transfer point (STP) and service control point (SCP)

(c) Service switching point (SSP) and service control point (SCP)

(d) Service switching point (SSP), signal transfer point (STP), and service control point (SCP)

3.10 The SS7 signaling connection control part (SCCP) is the layer that is responsible for services that are connected oriented, connectionless, and peer-to-peer communication which includes the establishment of connection, the starting of data transfer, and the releases of the connections.

(a) True

(b) False

Answer: 3.1 b, 3.2 a, 3.3 b, 3.4 a, 3.5 e, 3.6 c, 3.7 a, 3.8 b, 3.9 d, 3.10 a

Problems

3.1 Name the seven layers of the OSI model.

3.2 Describe the physical layer and the application layer.

3.3 (a) How many layers does the TCP/IP model have? (b) Describe each of the models.

3.4 What is a hub?

3.5 Describe the two different types of hubs.

3.6 Describe the different types of bridges.

3.7 What are routers, and what are the two ways through which routers can receive information?

3.8 Describe Signaling System No. 7 (SS7).

3.9 Name the three types of signaling points and describe each of them.

3.10 Name and describe the SS7 layer models.

3.11 What is a repeater designed to do?

3.12 Explain a gateway.

3.13 Describe a brouter.

3.14 Compare the OSI model with the IEEE model.

Local Area Networks

4

Networking is not about just connecting people. It's about connecting people with people, people with ideas, and people with opportunities.

Michele Jennae

Abstract

A local area network (LAN) connects personal computers, workstations, printers, servers, and other devices. There are different topologies (bus, ring, star, tree, etc.), transmission medium, and medium access control (MAC) that are included in the architectures of a LAN. This chapter covers LANs and the different types of LANs. The advantages and disadvantages of the different topologies of LAN are covered. All of the operational access methodologies including the controlled access devices are also discussed.

Keywords

Local area network (LAN) · IEEE 802 standards · Carrier Sense Multiple Access with Collision Avoidance (CSMA/CA) · Medium access control (MAC)

4.1 Introduction

Local area network (LAN) is typically a group of data communication networks or a group of computers and other necessary devices that are connected within limited geographical areas such as a building or a campus. A typical LAN is shown in Fig. 4.1. Its origin could be traced to the introduction of the IBM terminal equipment in 1974. A LAN connects personal computers, workstations, printers, servers, and other devices. It is governed by IEEE 802 standards that define the layers which in most cases are the data link and physical layers. Characteristically, LANs use packet *broadcasting*. This is a term that refers to transmissions that are available to large audiences. The costs are relatively low and the speeds are high.

A **local area network** connects computers within a limited geographical areas and are almost always belong to one organization.

The LAN architectures include different topologies (bus, ring, star, tree, etc.), transmission medium, and medium access control (MAC). The applications of LAN include personal computer LANs, backbone LANs (interconnect low-speed local LANs), high-speed office networks, back end networks (interconnecting large systems), and storage area networks (network handling storage needs). LANs use different access techniques that include random access and controlled access protocols.

This chapter covers LANs and the different types of LANs. The advantages and disadvantages of the different topologies of LAN are covered. All of the operational access methodologies

M. N. O. Sadiku, C. M. Akujuobi, *Fundamentals of Computer Networks*,
https://doi.org/10.1007/978-3-031-09417-0_4

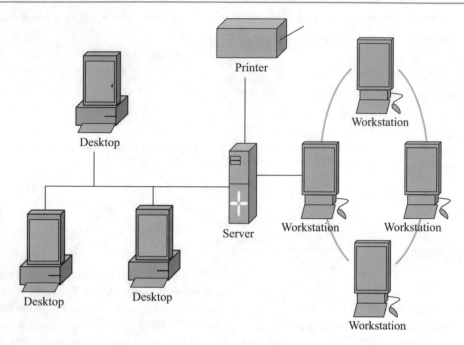

Fig. 4.1 Typical local area network

including the controlled access devices are also discussed.

4.2   Types of LANs

Local area network (LAN) has many topologies. The common types of LAN topologies are star, token ring, bus, and tree.

4.2.1   Star LAN Topology

A star topology is a network that is designed to look very similar to a star with a central core and many devices connected directly to that core as shown in Fig. 4.2. The systems in a star topology do not connect to each other but instead pass messages to the central core and in turn passes the message to either all other systems (or devices) or the specific destination system (or devices) depending on the network design. This topology works well for many smaller networks and works around many of the detriments associated with bus or ring topologies. You can see the general design of this topology in Fig. 4.2.

A star topology does have its own limitations, but there are effective ways of working around them. In reality, you can only connect to so many systems to the same star network before you begin to run into physical limitations, such as cable length or the number of ports available on the hardware used for the network. The star topology handles this by being easily extended into multiple stars with a central core in the middle.

In a star topology, computers are not connected to one another but are all connected to a central hub or switch. When a computer sends data to the other computers on the network, it is sent along the cable to a central hub or switch, which then determines which port it needs to send the data through for it to reach the proper destination. Characteristics of a star topology are as follows:

- A star topology is scalable.
- All cables run to a central connection point.
- If one cable breaks or fails, only the computer that is connected to that cable is unable to use the network.
- Because there is so much cabling used to connect individual computers to a central point,

Fig. 4.2 Diagram of a
star LAN topology.
(Source: www.en.
wikipedia.org)

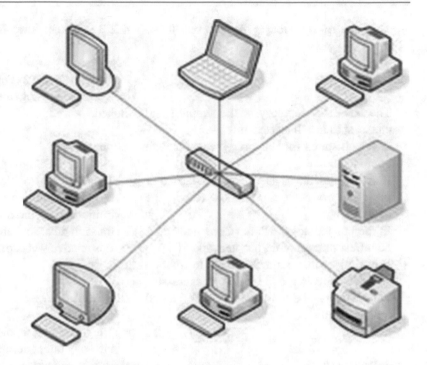

this may increase the cost of expanding and maintaining the network.

- As the network grows or changes, computers are simply added or removed from the central connection point, which is usually a hub or a switch.

Advantages

- The star topology is the most reliable because the failure of a node or a node cable does not affect other nodes.
- It is simple to troubleshoot because only one node is affected by a cable break between the switch and the node.
- Adding new nodes to the star topology does not greatly affect performance because the data does not pass through unnecessary nodes.
- The star topology can easily be upgraded from a hub to a switch or with a higher performance switch.
- It is also easy to install and to expand with extra nodes.

Disadvantages

- The star topology uses the most cable which makes it more expensive to install than the other topologies, especially the token ring and the bus topologies.
- The extra hardware required to implement the star topology such as hubs or switches increases the cost more.
- In the star topology, as the central computer controls the whole system, the whole system will be affected if it breaks down or if the cable link between it and the switch fails.
- Also, if the switch, the link to the server, or the server itself fails, then the whole network fails.

4.2.2   The Bus LAN Topology

In the bus LAN topology, the nodes are connected to a bus cable as shown in Fig. 4.3. If data is being sent between nodes, then other nodes cannot transmit. If too many nodes are connected, then the transfer of data slows dramatically as the nodes have to wait longer for the bus to be clear.

Advantages

- The bus LAN topology is the simplest and cheapest to install and extend.
- It is well suited for temporary networks with not many nodes.
- It is very flexible as nodes can be attached or detached without disturbing the rest of the network.
- In the bus topology, failure of one node does not affect the rest of the bus network.
- It is simpler than a ring topology to troubleshoot if there is a cable failure because sections can be isolated and tested independently.

Disadvantages

- In the bus topology, if the bus cable fails, then the whole network will fail.
- The performance of the network slows down rapidly with more nodes or heavy network traffic.
- The bus cable has a limited length and must be terminated properly at both ends to prevent reflected signals.
- It is slower than a ring network as data cannot be transmitted while the bus is in use by other nodes.

4.2.3 Token Ring Topology

In a token ring topology, the nodes are connected in a ring as shown in Fig. 4.4, and data travels in one direction using a control signal called a "token."

Advantages

- The token ring topology is not greatly affected by adding further nodes or heavy network traffic as only the node with the "token" can transmit data, so there are no data collisions.
- It is relatively cheap to install and expand.

Disadvantages

- The token ring topology is slower than a star topology under normal load.
- In the token ring topology, if the cable fails anywhere in the ring, then the whole network will fail.
- The token cannot be passed around the ring any longer in its operation if any node fails, which then results in the whole network failing.
- It is the hardest topology to troubleshoot because it can be hard to track down where in the ring the failure has occurred.
- The token ring is harder to modify or expand because to add or remove a node, you must shut down the network temporarily.

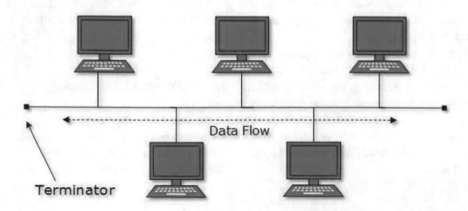

Fig. 4.3 Bus network topology. (Source: www.medium.com)

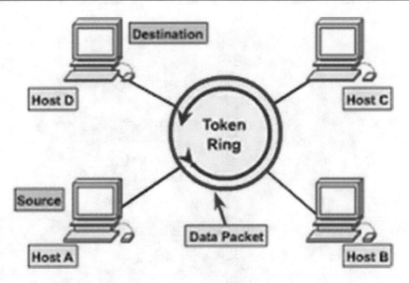

Fig. 4.4 Diagram of a token ring LAN. (Source: https://www.pinterest.com/pin/828662400161615931/)

- In the token ring topology operation, in order for the nodes to communicate with each other, they must all be switched on.

4.2.4 LAN Tree Topology

A tree topology combines characteristics of linear bus and star topologies. It consists of groups of star-configured workstations connected to a linear bus backbone cable as shown in Fig. 4.5. Tree topologies allow for the expansion of an existing network. They also enable schools to configure a network to meet their needs.

Advantages

- The tree topology has point-to-point wiring for individual segments.
- It is supported by several hardware and software vendors.

Disadvantages

- In a tree topology, the overall length of each segment is limited by the type of cabling used.

- It is more difficult to configure and wire than other topologies.
- In a tree topology, the entire segment goes down if the backbone line breaks.

4.3 Random Access

Random access is important because any storage location can be accessed directly. It is also called a direct access storage or memory. It is organized and controlled in a way that enables data to be stored and retrieved directly to specific locations. Random access assigns data resources dynamically to a large set of users, each with relatively busty traffic. Random access devices are used to allow data items to be read or written in almost the same amount of time irrespective of the physical location of data inside the memory. It is also called contention methods.

The ability to access data at random is called *random access*. It is the ability to access an arbitrary element of a sequence in equal time or any datum from a population of addressable elements roughly as easily and efficiently as any other, no matter how many elements may be in the set.

Some of the examples of the random access protocols are ALOHA, Carrier Sense Multiple Access (CSMA), Carrier Sense Multiple Access/Collision Detection (CSMA/CD), and Carrier

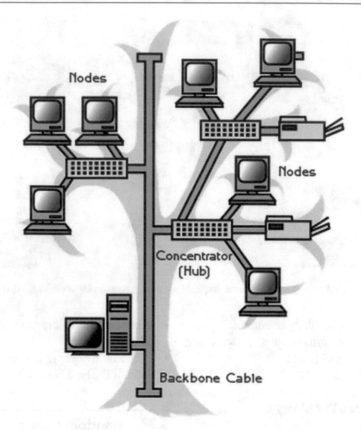

Fig. 4.5 LAN tree topology. (Source: https://computernetworkingtopics.weebly.com/uploads/1/0/2/3/10235412/1250759. gif?433)

Sense Multiple Access with Collision Avoidance (CSMA/CA). No station is superior to another station, and none is assigned control over another in random access or contention methods. In addition, no station permits, or does not permit, another station to send. At each instance, a station that has data to send uses a procedure defined by the protocol to make a decision on whether or not to send. The decision also depends on the availability of the medium (idle or busy).

4.3.1   ALOHA

The ALOHA protocol was developed in the early 1970s at the University of Hawaii to connect computers situated on different Hawaiian Islands. The computers of the ALOHA network transmit on the same radio channel whenever they have a packet to transmit. In fact, ALOHA

may be regarded as the father of multiple access protocols. The ALOHA protocol is a suitable protocol for satellite and terrestrial radio transmissions.

There are two types of ALOHA protocols, namely, slotted and unslotted ALOHA. In the slotted ALOHA, all frames have the same size, and the time is divided into equal size slots (time to transmit 1 frame). The nodes start to transmit and only slot beginning nodes are synchronized. However, if two or more nodes are transmitted in the slot, all nodes will detect collision. Operationally, when a node obtains fresh frame, it transmits it in the next slot. However, if there is no collision, the node can send the new frame in the next slot. Otherwise, if there is collision, the node retransmits the frame in each subsequent slot with a probability p until success is achieved. The ALOHA has a single active node that can continuously transmit at full rate of the channel.

It is highly decentralized and slots are normally synchronized. However, in ALOHA systems, collisions are experienced and slots are wasted. There are idle slots and nodes that may be able to detect collision in less than the time to transmit the packet. The clock is synchronized.

In the case of the unslotted Aloha, also called *Pure ALOHA*, it is simpler than the slotted ALOHA, and there is no synchronization. The packet needs transmission which means the packet can be sent without awaiting for the beginning of the slot. There is always the tendency for collision probability to increase, which means that the packet sent at $t0$ will collide with other packets sent in $t0-1$, and $t0+1$.

Example 4.1
Suppose you have N stations that have packets to send. Each slot transmits with a probability of p. What is the maximum probability of achieving a successful (S) transmission?

Solution:

Let N be the number of stations that have packets to send.

Let p be the probability of transmission in each slot.

Let the probability of successful transmission be S.

Using each single node in the slotted ALOHA, $S = p (1-p) (N-1)$ by any of N nodes.

This implies that S = Probability (only one transmits) = N p $(1-p)$ $(N-1)$.

Therefore, choosing optimum p as n -> infinity = $1/e$ = .37 as N -> infinity.

Example 4.2
Determine the probability that a node has been transmitted successfully in a pure Aloha.

Solution:

P(success by given node) = P(node transmits) P(no other node transmits in $[p_0-1, p_0]$ = p . $(1-p).(1-p)$

P(success by any of N nodes) = Np . $(1-p)$. $(1-p)$

Therefore, choosing optimum p as n -> infinity = $1/(2e)$ = .18

4.3.2 CSMA

Carrier Sense Multiple Access (CSMA) is a carrier sensing protocol mechanism used in random access schemes. It is used to find out if the system channel is busy, then it waits and proceeds to transmit when it is free. If there should be collision detected during the transmission of information, it will wait and retransmit the information frame later when the channel is free. The CSMA technique relies on the sender sensing the state of the transmission channel and basing its actions on this. It can therefore only be effectively used in channels which have short propagation delays, since for channels with long delays such as satellite channels, the sensed data will be considerably out of date. There are three types of CSMA protocols: 1-persistent CSMA, non-persistent CSMA, and p-persistent CSMA protocols.

1-Persistent CSMA
The 1-persistent CSMA is an aggressive version of the Carrier Sense Multiple Access (CMSA) protocol that operates in the medium access control (MAC) layer. When a transmitting station has a frame to send, the entire packet is transmitted if it senses that the channel is idle. If it senses that the channel is busy, it will wait and then retransmit as soon as the channel becomes free.

All users listen to the line prior to transmitting in the 1-persistent CSMA. If traffic is sensed, they wait. As soon as the line becomes free, the users that have been waiting will immediately transmit with probability one. If two stations start transmission simultaneously, collisions will occur. If the delay in the transmission channel is such that after one channel has started transmission the second ready channel has not heard this and also starts transmission, collisions will also occur. The stations wait a random amount of time before listening to the channel again after a collision.

If all transmitted packets are assumed to be of constant length and the load (G) has a Poisson distribution, then the throughput S is given by Eq. 4.1. It is also assumed in this calculation that all users can sense the transmission of all other users. The channel can be assumed to be overloaded if the throughput S is greater than load G, and if the throughput S is less than the load G, it is underloaded.

$$S = \frac{G_e^{-G}e(1+G)}{G+e^{-G}} \qquad (4.1)$$

where e^{-G} is the probability of a single transmission during a slot time.

Whether it is a slotted channel or unslotted channel, Eq. 4.1 holds. The assumption here is that for the slotted channels, the slot time will be much shorter than the packet time.

Example 4.3
A slotted ALOHA channel has an average 20% of the slots idle.

(a) What is the offered traffic G?
(b) What is the throughput?
(c) Is the channel overloaded or underloaded?

Solution:

(a) Probability of a single transmission during a slot time is e^{-G}.
   $20\% = e^{-G}$
   $0.2 = e^{-G}$
   This implies that G = −1.61.
(b) $S = G * e^{-G}$
   $= 1.61 * e^{-(1.61)}$
   $= 1.61*0.2$
   $= 0.32$
(c) For slotted ALOHA, S is maximum at $G = 1$.

In this case, G = 1.61 and S = 0.32. This means that it is beyond G = 1; hence, it is overloaded.

Non-persistent CSMA
In the non-persistent CSMA, if the channel is sensed idle, it will transmit the entire packet. However, if it is sensed busy, it will wait ran-

domly before it tries again. When there is collision, it will wait a random time and then restart the process again. The non-persistent CSMA protocol uses these two steps:

(i) The first step is when the channel is "sensed." If it is found to be idle, the packet is transmitted immediately.
(ii) In the second step, if the channel is "sensed" busy, the packet transmission is rescheduled using the retransmission distribution. At the end of the delay period, the protocol moves to step (i).

P-persistent CSMA
The p-persistent CSMA protocol when compared with 1-persistent CSMA and non-persistent CSMA protocols takes a moderate approach between them. It tells you the value of the probability p of the transmission after detecting that the channel is idle. The station first checks if the channel is idle and then transmits a frame with the probability p if it is idle. However, with a probability of $q=1-p$, it will wait for the next slot and then try again with a probability of p. If the channel sensed it is busy, it will wait until the next slot and then try again. When there is collision, it will wait for a random time and then restart the process. In general, at the heavier load, decreasing p would reduce the number of collisions. At the lighter load, increasing p would avoid the delay and improve the utilization. The value of p can be dynamically adjusted based on the traffic load of the network.

4.3.3   CSMA/CD

In the other CSMA systems, once transmission commences, it is completed even if a collision occurs on the first bit. Bandwidth is therefore wasted with such operation. The Carrier Sense Multiple Access with Collision Detection (CSMA/CD) is a way to solve the bandwidth waste problem. With the CSMA/CD protocol, a user monitors the line even when it commences transmission and stops once a collision is detected. After that, the user waits a random time before sensing the line again.

Collisions are detected in a short time in the CSMA/CD. Collision occurs when two stations listen for network traffic, and when they hear nothing, they transmit simultaneously resulting in damaging either transmissions or collision. In this case, the colliding transmissions are aborted reducing channel wastage. In collision detection mode, measuring signal strengths and comparing transmitted and received signals are easy in wired LANs. In the wireless LANs which are covered in Chap. 10, it is difficult if the receiver shuts off while transmitting. Collision detection enables stations to detect collisions, so they know when they must transmit. CSMD/CD is used by Ethernet LANs. The CSMA/CD was introduced by Xerox early in the 1970s. It is governed by IEEE 802.3 standard developed in 1985. The stations access the medium randomly. In addition, they contest for time on the medium.

It is also not always possible for a sender to effectively sense the line during its transmission since the strength of the transmitted signal may be so strong as to swamp any signal returned back. Often CSMA/CD algorithms require that each user which detects a collision transmit a short jamming signal to immediately inform all other users that a collision has occurred on the line. The transmission channel for the CDMA/CD may be slotted or unslotted irrespective of the variations in the protocols.

Example 4.4
A networking system using CSMA/CD has a bandwidth of 15 Mbps. Suppose the maximum propagation time including the delays in the devices and ignoring the time needed to send a jamming signal is 20 μs, what is minimum size of the frame?

Solution:

Let Tp = 20 μs and let the size of the bandwidth = 15 Mbps.

Therefore, the transmission time for the frame is Tfr = 2 × Tp = 40 μs. This implies that a station needs to transmit in the worst-case scenario for a period of 40 μs to detect the collision. In calculating the minimum size of the frame, it will be 15 Mbps × 40 μs = 600 bits or 75 bytes.

Ethernet
The Ethernet is a network access method that originated from the University of Hawaii and later adopted by Xerox Corporation. It is also called a media access method. It is governed by the IEEE 802.3 standard starting from the early 1980s. Ethernet is the most pervasive network access method in use today. It is the most commonly implemented media access method in LANs. Ethernet was the first network to provide CSMA/CD. It was developed in 1976 by Xerox Palo Alto Research Center (PARC) in cooperation with DEC and Intel corporations. Ethernet as shown in Fig. 4.6 is a fast and reliable network solution. It is one of the most widely implemented LAN standards. It can provide speeds in the range of 10 Mbps to 10 Gbps. It is used with a bus or star topology.

The different types of Ethernet LANs are:

- 10Base-T
  - It operates at 10 Mbps and governed by IEEE 802.3 standard.
- Fast Ethernet (100Base-T)
  - It operates at 100 Mbps and uses twisted pair cables.
- Gigabit Ethernet
  - It operates at 1 Gbps and uses fiber-optic cable.
- 10 Gbps Ethernet
  - This is the latest development of Ethernet models. It uses fiber-optic cable and is developed to meet the increasing bandwidth needs of LAN demands.
- Wireless Ethernet

  - It operates at 2.4 Gbps and is governed by the IEEE 802.11 standard.

4.3.4   CSMA/CA

In Carrier Sense Multiple Access with Collision Avoidance (CSMA/CA), if the station finds the channel busy, it does not restart the timer of the contention window; it stops the timer and restarts

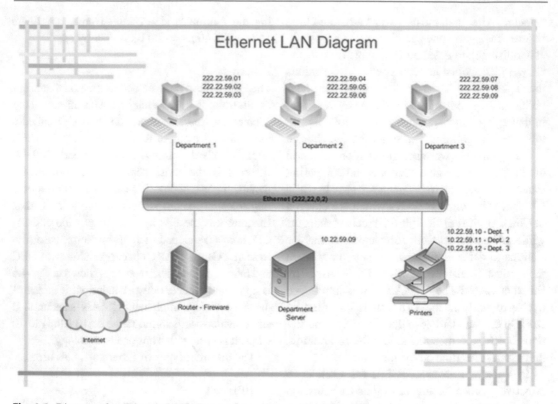

Fig. 4.6 Diagram of an Ethernet LAN. (Source: https://www.edrawsoft.com)

it when the channel becomes idle. Similar to CSMA, instead of sending packets, control frames are exchanged. The sender sends an RTS = request to send, the receiver sends a CTS = clear to send frame, actual DATA = actual packet is sent, sender transmits the data, and receiver responds with an ACK = acknowledgement frame, ensuring reliable transmission. As part of the advantages of the CSMA/CA, small control frames lessen the cost of collisions (when data is large), and RTS and CTS frames provide "virtual" carrier sense which protects against hidden terminal collisions (where terminal A cannot hear terminal B). However, as part of the disadvantages of the CSMA/CA, it is not as efficient as CSMA/CD. It is used by 802.11 wireless LANs.

4.4   Controlled Access

There are two main access control techniques that exit for computers to communicate with each other over the network. They are token ring-based access and token bus-based access.

4.4.1   Token Ring-Based Access

A frame, called a token, travels around the ring and stops at each node as shown in Fig. 4.4. If the node wants to transmit data, it adds that data and the addressing information to the frame. The advantage of using this method is that there are no collisions of data packets. When there is a single ring, all the devices on the network share a single cable, and the data travels in one direction only as shown in Fig. 4.4. When there are dual rings, two rings allow data to be sent in both directions. This helps to create redundancy (fault tolerance), meaning that in the event of a failure of one ring, data will still be transmitted on the other ring. The token ring-based access technique is governed by the IEEE 802.5 standard. Cambridge Ring that was developed at the University of Cambridge is an example of a token ring.

The token-based controlled access is used in ring network topologies (token ring) as shown in Fig. 4.4. In this case, each computer in the network can only transmit its data if it has the token.

This prevents collisions that occur when data is sent at the same time over the network or when it is randomly accessed. The token is a special pattern of bits/bit in a frame like 11111111 that is directly detectable by each node in the network. A computer may only transmit information if it is in possession of the token. The message is sent to all other computers in the network.

4.4.2   Token Bus Access

A token bus is governed by IEEE 802.4. The physical media is a bus as shown in Fig. 4.3 or a tree and a logical ring created using coaxial cable. The token is passed from one user to the other in a sequence that is either clockwise or anticlockwise. The stations know the addresses in either left or right directions as it regards the sequence in the logical ring. In terms of access to transmitting data, no station transmits data unless it has the token. This idea is also similar to the token ring. The token bus was created by IBM to connect their terminals to IBM mainframes. It is a 4 Mbps local area networking technology. The token bus uses a copper coaxial cable to connect multiple end stations such as terminals, workstations, and shared printers to the mainframe. The coaxial cable serves as a common communication bus and a token created by the token bus protocol to manage access to the bus. Any station that holds the token packet has permission to transmit data. The station releases the token when it is done communicating or when a higher priority device needs to transmit such as the mainframe. This keeps two or more devices from transmitting information on the bus at the same time and accidentally destroying the transmitted data.

However, the token bus has two major limitations. Failure in the bus also causes all the other devices beyond the failure to be unable to communicate with the rest of the network. Secondly, token bus does not allow easily the adding of more stations to the bus. New stations that are improperly attached are most unlikely to be able to communicate, and all devices beyond them are also affected. Thus, token bus networks are seen as somewhat unreliable and difficult to expand and upgrade.

Summary

1. A LAN connects personal computers, workstations, printers, servers, and other devices.
2. The systems in a star topology do not connect to each other but instead pass messages to the central core and in turn passes the message to either all other systems (or devices) or the specific destination system (or devices) depending on the network design.
3. A star topology does have its own limitations, but there are effective ways of working around them. In reality, you can only connect to so many systems to the same star network before you begin to run into physical limitations, such as cable length or the number of ports available on the hardware used for the network.
4. In the bus LAN topology, the nodes are connected to a bus cable.
5. In the slotted ALOHA, all frames have the same size, and the time is divided into equal size slots (time to transmit 1 frame).
6. In the case of the unslotted ALOHA, also called pure ALOHA, it is simpler than the slotted ALOHA, and there is no synchronization.
7. In CSMA protocol, you have 1-persistent CSMA, non-persistent CSMA, and p-persistent CSMA.
8. In the CSMA/CD, collisions are detected in a short time. Collision occurs when two stations listen for network traffic, and when they hear nothing, they transmit simultaneously resulting in damaging both transmissions.
9. In CSMA/CA, if the station finds the channel busy, it does not restart the timer of the contention window; it stops the timer and restarts it when the channel becomes idle. Similar to CSMA, instead of sending packets, control frames are exchanged.
10. The Carrier Sense Multiple Access/Collision Detection (CSMA/CD) that is part of the control access protocol is usually used in a bus topology.
11. In a token ring, when there is a single ring, all the devices on the network share a single cable, and the data travels in one direction

only, and when there are dual rings, two rings allow data to be sent in both directions.

12. Ethernet is the most pervasive network access method in use today. It is the most commonly implemented media access method in LANs.

13. The token-based controlled access is used in a ring network topologies (token ring).

14. The token bus was created by IBM to connect their terminals to IBM mainframes.

Review Questions

4.1 Local area network (LAN) is typically a group of data communication networks or a group of computers and other necessary devices that are connected within the same location/space.
(a) True
(b) False

4.2 Examples of the random access protocols include the following:
(a) ALOHA
(b) Carrier Sense Multiple Access (CSMA)
(c) Carrier Sense Multiple Access/ Collision Detection (CSMA/CD)
(d) Carrier Sense Multiple Access with Collision Avoidance (CSMA/CA)
(e) All of the above

4.3 ALOHA is not highly decentralized and slots are not normally synchronized.
(a) True
(b) False

4.4 In CSMA protocol, the following:
(a) 1-persistent CSMA
(b) Non-persistent CSMA
(c) p-persistent CSMA
(d) All of the above

4.5 CSMA/CD is governed by
(a) IEEE 802.2 standard
(b) IEEE 802.4 standard
(c) IEEE 802.3 standard
(d) IEEE 802.1 standard

4.6 In CSMA/CA, if the station finds the channel busy, it does not restart the timer of the contention window; it stops the timer and restarts it when the channel becomes idle.
(a) True
(b) False

4.7 The Carrier Sense Multiple Access/ Collision Detection (CSMA/CD) is not the part of the control access protocol that is usually used in a bus topology.
(a) True
(b) False

4.8 Ethernet can provide speeds in the range of:
(a) 1 Mbps to 10 Gbps
(b) 5 Mbps to 10 Gbps
(c) 15 Mbps to 10 Gbps
(d) 10 Mbps to 10 Gbps

4.9 Token bus networks are seen as somewhat unreliable and difficult to expand and upgrade.
(a) True
(b) False

4.10 In a star topology,
(a) Computers are not connected to one another but are all not connected to a central hub or switch.
(b) Computers are connected to one another but are all connected to a central hub or switch.
(c) Computers are not connected to one another but are all connected to a central hub or switch.
(d) Computers are connected to one another but are all not connected to a central hub or switch.

Answer: 4.1 a, 4.2 e, 4.3 b, 4.4 d, 4.5 c, 4.6 a, 4.7 b, 4.8 d, 4.9 a, 4.10 c

Problems

4.1 (a) What do you understand by random access relative to a LAN? (b) Describe briefly random access in a LAN.

4.2 Suppose you have three stations that have packets to send. Each slot transmits with a probability of .25. What is the maximum probability of achieving a successful (S) transmission?

4.3 Describe briefly Carrier Sense Multiple Access (CSMA) protocol in a random access LAN.

4.4 A networking system using CSMA/CD has a bandwidth of 30 Mbps. Suppose the maximum propagation time including the delays in the devices and ignoring the time needed to send a jamming signal is 40 μs, what is minimum size of the frame?

4.5 Describe briefly Carrier Sense Multiple Access with Collision Avoidance (CSMA/CA).

4.6 What is a token ring?

4.7 Describe briefly the different types of Ethernet LAN.

4.8 Mention some of the characteristics of a star topology in a LAN.

4.9 What are the characteristics of a LAN?

4.10 Why is Ethernet the most popular LAN? (Include the answer in the next)

4.11 Describe the token passing scheme in token ring.

4.12 What are the limitations of star LAN?

4.13 Mention the characteristics of a star topology.

4.14 What are the advantages of a star topology?

4.15 What are the disadvantages of a star topology?

4.16 What are the advantages of the bus LAN topology?

4.17 What are the disadvantages of the bus LAN topology?

4.18 What are the advantages of the token ring topology?

4.19 What are the disadvantages of the bus LAN topology?

4.20 What are the advantages of the LAN tree topology?

4.21 What are the disadvantages of the LAN tree topology?

4.22 A slotted ALOHA channel has an average 20% of the slots idle.
(a) What is the offered traffic G?
(b) What is the throughput?
(c) Is the channel overloaded or underloaded?

The Internet

5

The Internet is a testament to a connected system that works – it's a global network where any computer can reach another, and easily transfer information across.

John Collison

Abstract

This chapter discusses the Internet, which is a global system of interconnected computer networks that use the Internet protocol suite to link devices worldwide. The Internet is a widespread information infrastructure, which is often called the information superhighway. Typically, the Internet uses TCP/IP protocols in combination with other devices such as routers, switches, modem, and firewalls. TCT/IP provides users with services such as emailing, file transfer, remote login, and web browsing. The Internet has made a tremendously impact on our culture, commerce, education, and government. It's applications include Internet telephony, Internet radio, Internet television, manufacturing, social networking, entertainment, gamification, education, e-commerce, stock trading, digital libraries, geographic information systems, and government.

Keywords

Internet protocol · Internet protocol · TCP/IP protocols · Emailing · File transfer · Remote login · Web browsing

5.1 Introduction

The Internet is a global system of interconnected computer networks that use the Internet protocol suite to link devices worldwide. It may also be regarded as a network of networks consisting of private, public, academic, business, and government networks. The Internet is a combination of networks, including the ARPANET, NSFnet, regional networks such as NYsernet, local networks at a number of universities and research institutions, and a number of military networks. Being the largest wide area network, the Internet is good for long-distance communications like X.25 and frame relay.

> The **Internet** is a global network (or wide area network) comprising several interconnected autonomous networks.

The Internet today is a widespread information infrastructure, which is often called the information superhighway. Any computer connected to the Internet is regarded as a host. Internet service providers (ISPs) provide the end system access to the Internet. The ease of access provided by the Internet has shrunk our world, opened international borders, and allowed people to share information in an unprecedented way. Businesses of every size and kind, education systems, governments, and organizations embrace the Internet as a means of sharing information and resources, conducting commerce, and engaging socially.

In this chapter, we first introduce TCP/IP protocols and IP addresses. We consider some specific Internet applications and services. We discuss important issues facing the Internet: privacy, security, and safety. We finally cover the next-generation Internet (IPv6) and Internet2, which is the future of Internet.

5.2 Protocol Suite

The popularity of the Internet gained momentum with the advent of the World Wide Web in 1991. The next technological step was when mobile devices were able to connect to the Internet. Like LANs, the Internet is a packet-switching network. The data or information is divided into packets with headers containing source and destination addresses.

5.2.1 TCP/IP Protocols

All computers connected to the Internet do not speak the same language, but if they are going to be networked, they must share a common set of rules known as *protocols*. A protocol is roughly a set of rules for communication. The core technology supporting the Internet is called Transmission Control Protocol/Internet Protocol, or TCP/IP. TCP/IP was developed by Vinton Cerf (the father of the Internet) along with his colleague Robert Kahn. Typically, the Internet uses TCP/IP protocols in combination with other devices such as routers, switches, modem, and firewalls. Using OSI terminology, TCP is a layer 4 protocol, while IP is a layer 3 protocol. All computers connected to the Internet run IP software; most of them also run TCP software.

As shown in Fig. 5.1, TCP/IP is a layered set of protocols. Internet applications generally use four layers: application layer, transport layer, Internet layer, and network layer.

- *Application layer*: This is where the end users interact with the Internet. Application programs that use the Internet reside here. Application programs implementing TCT/IP provide users with services such as emailing, file transfer, remote login, and web browsing.
- *Transport layer*: This layer provides end-to-end services that are independent of the structure of user data. In the Internet, there are two transport protocols: TCP and UDP. TCP (Transmission Control Protocol) is connection oriented, and each session begins with a connection setup procedure (the so-called three-way handshake). It provides services needed by many applications. TCP breaks data into packets or provides segmentation of long messages. It is also responsible for performing congestion control. UDP (User Datagram Protocol) provides connectionless services. Since

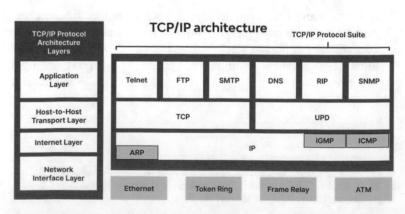

Fig. 5.1 Internet protocol architecture. (Source: "Transmission Control Protocol – TCP," https://www.wallarm.com/what/transmission-control-protocol-tcp)

the overhead of establishing a connection is eliminated, data can be transmitted faster using UDP. However, there is no flow control and no negative acknowledgment of damaged packets.

- *Internet layer*: This layer is responsible for addressing and routing of data. It breaks up large messages and reassembles them at the destination. It creates a data unit called a datagram. The Internet protocol (IP) moves datagrams from one host to another using various routing techniques and routers. It is an unreliable, connectionless protocol. IP contains four supporting protocols: ARP, RARP, ICMP, and IGMP. Address resolution protocol (ARP) is responsible for determining the address of each device on the network from its IP address.
- *Network layer*: This is the lowest component layer of the Internet protocols. As TCP/IP is designed to be hardware independent, TCP/IP may be implemented on top of virtually any hardware networking technology. Protocols at this layer manage the specific physical medium.

TCP/IP is an agreed upon standard for computer communication over the Internet. The protocols are implemented in software that runs on each node. The Internet Protocol (IP) can be regarded as the common thread that holds the entire Internet together. It is responsible for moving datagrams from one host to another, using various techniques (or "routing" algorithms).

An IP packet/datagram consists of a header followed by data (or payload). At a minimum, the header is 20 bytes long and with options can be significantly longer. The header format is shown in Fig. 5.2 and explained as follows.

- *Version* (4 bits): This is always 4 for IPv4.
- *IHL* (4 bits): The Internet Header Length (IHL) field specifies the length of the header, in 32-bit words. The minimum value for this field is 5 (indicating a length of 5×32 bits = 20 bytes), while the maximum value is 15 words (15×32 bits = 60 bytes).
- *DSCP* (6 bits): Differentiated Services Code Point (DSCP) is the type of service. New technologies such as VOIP are emerging that require real-time data streaming and therefore make use of the DSCP field.
- *ECN* (2bits): Explicit Congestion Notification field allows end-to-end notification of network congestion without dropping packets.
- *Total length* (16 bits): This field defines the entire packet size in bytes, including header and data. The minimum size is 20 bytes (header without data), and the maximum is 65,535 bytes.
- *Identification* (16 bits): This field is used for uniquely identifying the group of fragments of a single IP datagram.
- *Flags* (3 bits): A 3-bit field is used to control or identify fragments.
- *Fragment offset* (13 bits): The fragment offset field is measured in units of 8-byte blocks. It is 13 bits long and specifies the offset of a particular fragment relative to the beginning of the original unfragmented IP datagram. The

IPv4 Header Format

Offsets	Octet	0								1								2								3							
Octet	Bit	0	1	2	3	4	5	6	7	8	9	10	11	12	13	14	15	16	17	18	19	20	21	22	23	24	25	26	27	28	29	30	31
0	0	Version				IHL				DSCP						ECN		Total Length															
4	32	Identification																Flags			Fragment Offset												
8	64	Time To Live								Protocol								Header Checksum															
12	96	Source IP Address																															
16	128	Destination IP Address																															
20	160	Options (if IHL > 5)																															
24	192																																
28	224																																
32	256																																

Fig. 5.2 IP header format

first fragment has an offset of zero. This allows a maximum offset of $(2^{13} - 1) \times 8 = 65,528$ bytes, which would exceed the maximum IP packet length of 65,535 bytes with the header length included $(65,528 + 20 = 65,548$ bytes).

- *Time To Live* (TTL) (8 bits): This field (specified in seconds) helps prevent datagrams from wandering on the Internet. When the TTL field hits zero, the router discards the packet.
- *Protocol* (8 bits): This field defines the protocol used in the data portion of the IP packet.
- *Header Checksum* (16 bits): This field represents a value that is calculated using an algorithm covering all the fields in header. This field is used to check the integrity of an IP datagram.
- *Source address* (32 bits): This field specifies the IP address of the sender of the packet.
- *Destination address* (32 bits): This field specifies the IP of the intended receiver of the packet.
- *Options*: The options (variable length) field is not often used. This field represents a list of options that are active for a particular IP datagram. Options include security, record route, and time stamp. If the option values are not a multiple of 32 bits, 0s are added or padded to ensure this field contains a multiple of 32 bits.

5.2.2   IP Address

For IP to work, every computer must have its own number to identify itself. This number is called the IP address. We all know that computers like to work with numbers, and humans prefer *names*. With this in mind, the designers of the Internet have set up a system to give names to computers on the Internet. A domain name system (DNS) server is a computer somewhere that can change a *hostname* into an *IP address* and vice versa. Domain names usually end with names in Table 5.1. In the DNS naming of computers, there is a hierarchy of names as illustrated in Fig. 5.3.

To avoid addressing conflicts, IP addresses are assigned by the five regional Internet registries (RIRs) as shown in Fig. 5.4:

Table 5.1 Internet domains

Domain	Meaning
com	Commercial institution
cdu	Educational institution
gov	Government
int	International organizations
mil	US military
net	Network service provider
org	Nonprofit organization
tv	Television channel
ca	Canada
uk	United Kingdom
us	USA

- African Network Information Center (AfriNIC) for Africa
- American Registry for Internet Numbers (ARIN) for North America
- Asia-Pacific Network Information Centre (APNIC) for Asia and the Pacific region
- Latin American and Caribbean Internet Addresses Registry (LACNIC) for Latin America and the Caribbean region
- Réseaux IP Européens–Network Coordination Centre (RIPE NCC) for Europe, the Middle East, and Central Asia

Currently, the IPv4 protocol identifies each node through a 4-byte address. IP addresses are usually written as a sequence of four numbers separated by three dots such as NNN.NNN.HHH.HHH, where N stands for octets that identify network and H denotes octets that specify the host. This is known as dotted decimal notation. Each number can be between 0 and 255 except the last number which must be between 1 and 254. Typical IP addresses are 127.0.1.53 and 182.30.9.22. Let us break down the IP address "182.30.9.22" in Table 5.2. Table 5.3 presents some IP addresses and their DNS (domain name system) names.

There are five classes of available IP ranges: class A, class B, class C, class D, and class E. Each class allows for a range of valid IP addresses, as shown in Table 5.4.

Table 5.5 provides a list of valid and invalid IP addresses.

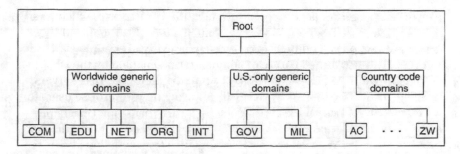

Fig. 5.3 DNS hierarchy. (Source: McDysan, 2000, p. 173)

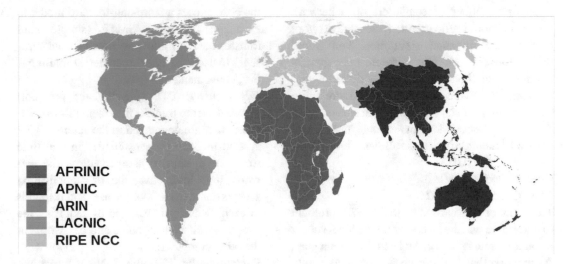

Fig. 5.4 The five regional internet registries. (Source: "Regional Internet registry," Wikipedia, the free encyclopedia, https://en.wikipedia.org/wiki/Regional_Internet_registry)

Table 5.2 IP address 182.30.9.22

IP	182	30	9	22
Binary value	10110110	00011110	00001001	00010110
Decimal value	128 + 32 + 16 + 4 + 2 = 182	16 + 8 + 4 + 2 = 30	8 + 1 = 9	16 + 4 + 2 = 22

Table 5.3 Example of IP addresses and their DNS names

IP address	DNS name
198.137.240.91	www.whitehouse.gov
18.72.0.3	bitsy.mit.edu
129.207.153.23	www.pvamu.edu
198.105.232.5	www.microsoft.com
198.95.249.66	ftp.netscape.com
255.255.255.255	Broadcast

Table 5.4 Classes of IP addresses

Class	Address range	Supports
Class A	1.0.0.1 to 126.255.255.254	Supports 16 million hosts on each of 127 networks
Class B	128.1.0.1 to 191.255.255.254	Supports 65,000 hosts on each of 16,000 networks
Class C	192.0.1.1 to 223.255.254.254	Supports 254 hosts on each of 2 million networks
Class D	224.0.0.0 to 239.255.255.255	Reserved for multicast groups
Class E	240.0.0.0 to 254.255.255.254	Reserved for future use or research and development purposes

Table 5.5 Examples of valid and invalid IP addresses

IP number	Valid?
2.3.4.5.6	No
6.6.6.6	Yes
11.32.4.0	No
5.300.24.50	No
255.255.255.255	Yes
125.284.40	No
132.108.111.86	Yes

As the number of computers on a network grows, network traffic grows, decreasing network performance. In such a situation, you would divide the network into different subnetworks and minimize the traffic across the different subnetworks. To divide a given network address into two or more subnets, you use subnet masks. Class A, B, and C networks have default masks, also known as natural masks, as follows:

Class A: 255.0.0.0

Class B: 255.255.0.0

Class C: 255.255.255.0

The subnet mask is used by the TCP/IP protocol to determine whether a host is on the local subnet or on a remote network. An IP address on a class A network that has not been subnetted would have an address/mask pair: 8.20.15.5 255.0.0.0. In order to see how the mask helps you identify the network and node parts of the address, convert the address and mask to binary numbers.

8.20.15.5   =

00001000.00010100.00001111.00000101

255.0.0.0 = 11111111.00000000.00000000.0000
0000

From this, any address bits which have corresponding mask bits set to 1 represent the network ID. Any address bits that have corresponding mask bits set to 0 represent the node ID.

5.2.3　Internet Services

Application programs implementing TCT/IP provides users with services such as emailing, file transfer, remote login, and web browsing.

- *Electronic mail*: Email (or computer-mediated communication) is perhaps the most prevalent application of the Internet. It allows people to write back and forth without worrying about the cost. It uses a store-and-forward model to exchange messages. SMTP (simple mail transfer service) allows a user to send messages to users on other computers. Emailing is fast, convenient, and free.
- *File transfer*: FTP (file transfer protocol) allows a user to transfer files to and from computers that are connected to the Internet. FTP is used to copy or transfer files in real time from one system to another across the network. It is interactive; the FTP application accepts commands from the user and responds to each. Security is handled by requiring the user to specify a user name and password for the other computer.
- *Remote login*: TELNET (network terminal protocol for remote login) allows a user to access or log on to and use other computers that are connected to the Internet regardless of their location. It allows you to enter commands into a remote computer just as if you are entering commands at your own computer.
- *World Wide Web* (WWW): The WWW (or the web) incorporates all of the Internet services. The current foundation on which the WWW functions is the programming language called HTML (HyperText Markup Language). The Hypertext Transfer Protocol (HTTP) is a

8.20.15.5 = 00001000.00010100.00001111.00000101
255.0.0.0 = 11111111.00000000.00000000.00000000

----

net id |　　host id

netid = 00001000 = 8
hostid = 00010100.00001111.00000101 = 20.15.5

TCP/IP protocol that is used to access a WWW or HTML documents or pages. A Web page is created by typing HTML tags. The eXtensible Markup Language (XML) is for structuring web-based documents. It allows storing and exchanging data across different applications and operating systems. It complements HTML, which focuses on how to display documents.

In addition to these services, we also have the following standard application protocols as part of the TCP/IP protocol suite:

- *DNS* (domain name system): This translates a human name to an IP address (forward DNS) or an IP address to a domain name (reverse DNS). DNS allows users to work with much more simple symbolic names. The address is obtained by looking it up in the DNS, a distributed database containing name-address mappings for each Internet domain name. For example, 198.137.240.91 is the IP address for www.whitehouse.gov BIND (Berkeley Internet Name Domain) is by far the most widely used DNS software on the Internet.
- *RTP* (Real-Time Protocol): This is a real-time transport protocol for multiple applications such as voice over IP, Internet radio, Internet TV, videoconferencing, and other multimedia applications. It defines a standardized packet format for delivering audio and video over the Internet. The RTP is a simple protocol that runs on top of UDP and therefore has "best effort" delivery. It is specified by IETF in RFC 3550 and RFC 3551.

Voice over IP is a phone service over the public Internet.

- *RTCP* (Real-time Transport Control Protocol): This is primarily used for collecting data on the efficiency and quality of the connection. The RTCP messages travel on the same route as RTP and report information such as latency, jitter, and packet loss.
- *ARP* (Address Resolution Protocol): This figures out the unique address of devices on the network from their IP addresses. It is used for mapping a network address (e.g., an IPv4

address) to a physical address like an Ethernet address (also named a MAC address). ARP was defined by RFC 826 in 1982 as a critical function in the Internet protocol suite.

5.3   Internet Applications

Potentialities offered by the Internet make possible the development of a huge number of applications and services. The Internet has made a tremendously impact on our culture, commerce, education, and government. It's applications include Internet telephony, Internet radio, Internet television, manufacturing, social networking, entertainment, gamification, education, e-commerce, stock trading, digital libraries, geographic information systems, and government. We will consider some of these applications.

5.3.1   VOIP

Voice over Internet Protocol (VOIP) or Internet telephony is one of the most prominent telecommunication services based on the Internet Protocol (IP). It is a technology that allows communication of voice over the public Internet, rather than via the public switched telephone network (PSTN). It enables the real-time transmission of voice signals as packetized data over "IP networks" that employ the Transmission Control Protocol (TCP), Real-Time Transport Protocol (RTP), User Datagram Protocol (UDP), and Internet Protocol (IP) suite.

This technology converts analog voice signals into digital packets of information. These packets are transmitted over the Internet, allowing for conversations to take place anywhere in the world. A typical VoIP configuration is shown in Fig. 5.5. A VOIP network consists of a switched LAN infrastructure with WAN links between its main campuses. Perhaps the most important component in a

Fig. 5.5 A typical VOIP configuration. (Source: Miloszewski, 2011)

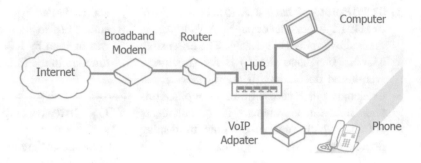

VOIP system is the gateway. The task of a gateway is to sit at the border of two different types of network and help them communicate. It is responsible for translating the information into a format that each network can understand. The network generally requires at least one C. VoIP "gateways" provide the bridge between the local PSTN and the IP network for both the originating and terminating sides of a call. To support standard telephony signaling, the gateways must employ a common protocol such as the H.323 or a proprietary protocol. H.323 is an ITU recommendation for multimedia communication over LANs. This standard is for point-to-point communications as well as multipoint conferences. It includes many protocols that provide call setup, call forwarding, management, and authorization as well as the transmission of data, voice, and video over a telephone call. The IEFT has defined the Session Initiation Protocol (SIP) which operates at the application layer and provides services such as call forwarding and signaling. The majority of the VoIP deployments use RTP (real-time protocol) for actual media transport since voice conversation is a real-time activity and no lag can be tolerated.

There are two major reasons for using VOIP: lower cost and increased functionality. VoIP has become a key enabling technology for multimedia communication on the IP network. It is being deployed rapidly. Major disadvantages are latency and unreliable transmission (loss of service during outages). Providers of VOIP services, such as Skype, Google Talk, Facebook Messenger, and WhatsApp, offer the benefit of free calls. These service providers see VOIP technology as a means of reducing their cost of offering existing voice-based services and new multimedia services. They have designed voice calls with the quality of the call not an issue. QoS (quality of service) is important for IP-based multimedia services. The signal quality of the VoIP system depends on several factors such as delay, jitter, packet loss, networking conditions, coding processes, and error correction schemes. The quality of VoIP can be significantly improved by using proper codec.

Although VOIP promises inclusion, innovation, and growth, it also brings some challenges. VoIP is not easy to secure due its interconnection to the PSTN. Unlike PSTN, there is no central entity responsible for the design, implementation, and monitoring of the voice services. VOIP lacks quality of service. Noisy lines and delays tend to frustrate users.

5.3.2   Social Networking

Social networks constitute the greatest global information platform on the Internet today. They have become an indispensable part of our daily lives as people spend more time socializing on the Internet.

A **social networking service** is an Internet-based platform used in building and developing social relations among people.

Major modern social networking sites include Facebook, Twitter, YouTube, and MySpace.

- *Facebook*: Facebook was first introduced in 2004 as a Harvard social networking site, expanding to other universities and eventually to everyone. It became the largest social networking site in 2009. It remains the largest photo sharing site. Marketing strategists have found Facebook to be useful because it covers a range of personal and organizational interests.
- *Twitter*: Twitter was founded in 2006 by Odeo, Inc. and was originally only for Odeo, Inc. employees and family members. It became a public network in 2006. Twitter provides a real-time, Web-based service which enables users to post brief messages for other users and to comment on other user posts. Tweets are extracted from Twitter. A tweet is a small message of no more than 140 characters that users create in order to communicate thoughts. Microblogging is a newer blog option made popular by Twitter.
- *YouTube*: This is a video sharing platform where many people can discover, watch, and share user-generated videos. It is a website of participatory culture. It has become the most successful Internet website providing a short video sharing ser-

Friendster, Vox. Bebo, LiveJournal, and Flickr. The impact of these modern social networks on various areas of our life has far surpassed the expectations of many. Common applications of social networking involve computer-mediated social interaction, education, business, finance, healthcare, politics, religion, and crowdsourcing.

5.3.3 Education

Digital or online education is the wave of the future in education. Education providers are moving from traditional face-to-face environments to those that are completely digital. Digital technologies (phone, computer, table, e-book, social networking, online videos, mobile devices, etc.) offer great hope for both learners and teachers today. The use of digital technologies to improve learning keeps climbing. Online K-12 schools are spreading across the USA. Schooling that combines computerized learning seems to be the emerging model. Our publics in digital education include students, educators, parents, administrators, policy-makers, and commercial interests.

Today, online education, based on new digital and networked technologies, is the fastest

Digital or online education is the process of using digital technology in teaching and learning.

vice since its establishment in early 2005. Since YouTube is a Google property, to sign up for a YouTube account requires a Google account.
- *MySpace*: This social networking site bases its existence on advertisers who are paying for page views. It has a lot that users could do. There are MySpace sites in the United Kingdom, Ireland, and Australia.

Others social networking sites include Instagram, Google+, LinkedIn, China-based Renren,

growing segment of higher education. Online education is needed to provide online learning to those who have difficulty attending the traditional university. Although online education may take many forms, it will not succeed except it provides the following services to distant students.

- *Administrative services*: Under these services, students should have access to admissions, initial advising, registration, scholarship, ongoing advisement, degree audit, course schedule, course catalog, financial aid, payment, transcript, and award of certification.

- *Electronic lectures*: Online education consists of virtual lectures, virtual tutorials, and virtual exams. It provides learning and education in the form of online courses through the Internet and multimedia technology. Electronic lectures used in distance education can be synchronous lectures or asynchronous lectures. They are used in presenting materials and can be transmitted through the Internet, cable, or satellite.
- *Digital library*: Students need access to digital libraries to obtain information. The resources that the digital libraries should provide include e-books, e-courses, e-journals, databases, and interlibrary loan services.

Online education applies the Internet and communication technologies and makes education open, dynamic, and affordable to those who want to learn, regardless of their age or location. For example, for-profit institutions such as University of Phoenix, Kaplan University, Walden University, and Athabasca University (in Canada) have dominated the online market. Now, there are evening classes, weekend classes, satellite campus, and cyberclasses.

Since the introduction of the first Massive Online Open Courses (MOOCs) in 2003 in the United Kingdom, millions of students across the

MOOCs are offered in many areas such as engineering, computer science, finance, business, education, health sciences, criminal justice, cybersecurity, IT, psychology, archeology, legal studies, and nursing. The main requirement is a computer system and Internet connectivity (cable, wireless, DSL, etc.).

Traditional education is facing a lot of challenges—becoming more expensive, shortage of professors, cut in funding, busy classrooms, course shortages, limited infrastructures, etc. Online education can solve some of these issues. It is scalable and less expensive. It allows students to work on the course anywhere there is Internet connection. There is no discrimination among students on the basis of race, sex, religion, and nationality.

5.3.4 E-Commerce

With the opening of the Internet for commercial activities in 1991, thousands of companies worldwide have started doing business online. To stay competitive, more and more companies are using the Internet to provide services online and increase market share. They conduct business transactions with their customers and vendors over the Internet.

It has revolutionized business transactions by

Electronic commerce (or e-commerce) refers to the process of conducting business transactions over the Internet.

world have grabbed the opportunity to take courses online. The MOOC phenomenon is recent, disruptive, and revolutionary in online higher education. MOOC is an open online course that allows anyone to register without time limitations, geographic restrictions, or prerequisites. The goal of MOOCs is to provide online education for busy people for the careers of tomorrow and extend knowledge and skills to the entire world. MOOC provides access and flexibility to students balancing work, family, and financial responsibilities. By definition, MOOCs are ·massive because they involve effectively teaching simultaneously thousands of students.

enabling the consumers to purchase, invest, bank, and communicate from virtually anytime, anywhere. For example, money can be transferred between banks electronically. Amazon and eBay are typical examples of e-commerce companies that rely on Internet-based technology to sell their products around the globe.

E-commerce has created opportunities for businesses to reach consumers directly. It facilitates cross-border transactions and brings about economic and social development. E-commerce has some advantages over traditional commerce: lower cost of running a store, no rent to pay, no barriers to time or distance, easier and more convenient to run a business, and increase of profits. Other benefits include around-the-clock availability, easy and convenient accessibility, speed of access, wider selection of products and services, and international reach.

Common applications related to e-commerce include online shopping, online banking, payment systems, electronic air tickets, hotel reservation, tourism, and teleconferencing. For example, the banking industry has persuaded consumers to adopt ATMs, telephone or home banking, and now online banking. The demand for online banking (also known as electronic banking or Internet banking) has increased worldwide. Online banking allows customers to conduct their banking transactions over the Internet. It supports online bill payment, electronic funds transfer, online shopping, and checkbook transactions. It also facilitates e-society and e-commerce particularly online shopping and online sales. Today online banking is a global phenomenon. It will continue to facilitate e-commerce and serve as a strong catalyst for economic development.

To start an online business, one must find a niche product that consumers cannot easily find

5.3.5 Government

The Internet is the mainstay of modern political life. It is making significant changes in all levels of government—federal, state, and local. Digital government (or networked government) refers to the use of information technologies such as the Internet and mobile computing to support government operations and provide government services. Digital government, also known as e-government, is a global phenomenon whereby public servants leverage information and communication technology (ICT) to better serve their constituents. ICT applications to government services can be divided into three broad categories: providing access to information, transaction services, and citizen participation. The ICT includes the Internet (and associated technologies such as email, the WWW, and social networking) and communication technologies such as mobile computing, cell phones, and global positioning systems.

The **digital government** seeks to enable people to access government information and services anywhere, anytime, on any device.

in traditional stores. Basic elements required to run a business online are e-commerce software, a payment processing service, and a merchant account. The e-commerce software enables one to build and maintain websites and associated databases that contain products and prices. Most online purchases are paid by credit cards, smart cards, electronic cash, or through a third-party such as PayPal. The merchant account refers to the business account into which the money from credit card purchases is deposited. In the USA, delivery companies such as UPS and FedEx provide door-to-door services to consumers.

The security risks involved in e-commerce have been a major concern. E-commerce provides security defenses such as firewalls, authentication scheme, and encryption. More about security will be discussed in Chap. 11.

Contemporary democracy is being facilitated and empowered by the Internet. It has influenced voting and induced social changes, unrest, uprisings, and revolutions all over the world. It will make government to be more accountable and enable citizens to exercise freedom of speech. It also helps to engage people in the democratic process and to get the younger generations involved in politics. It also enables the government to inform, involve, connect, and mobilize citizens through the Internet.

Demand for accurate election process has led governments worldwide to adopt electronic voting. This is the next stage in the evolution of democracy and part of the digital government initiatives. Electronic voting (EV) systems are being deployed worldwide to save money, increase convenience, increase voter turnout, speed up the tallying process, eliminate counting errors, and ease accessibility for disabled voters. Most democratic governments in the world use electronic voting to elect their leaders. EV was introduced in the USA in 1975. Estonia was the first country to allow its citizens to vote over the Internet in 2005.

The Internet is also a major external driver of open government. The Open Government (OG) movement started in January 21, 2009, when President Barack Obama issued a Memorandum on Transparency and Open Government. He developed and centered the concept of OG on three pillars of transparency, participation, and collaboration in government. Transparency involves transparency of government operations and transparency of government-held data. Governments can use the Internet and social media to share ideas with citizens and involve them in the policy-making process. Collaboration (or accountability) aims at involving all stakeholders in government operations and decision-making. It is the joint effort to participate in the democratic process. It includes free and open access to government information. Making government information available publicly has been termed a civil right by the United Nations. Open government has increased in importance and become an essential global agenda.

Other applications of Internet are intranets and extranets discussed in the next chapter. Emerging applications of the Internet such as Internet of Things (IoT), big data, cloud computing, and smart cities will be covered in Chap. 12.

5.4   Privacy, Security, and Safety

As the use of the Internet expands, so is the risk. The issues of privacy, security, and safety are associated with the various uses of the Internet. Internet users are vulnerable to privacy violations, security threats, and unsafe environment due a range of risky behaviors in the cyber world.

5.4.1   Privacy

The concept of privacy is deeply rooted in modern civilizations. New technologies are making it easier for governments and corporations to monitor our online activities like never before. These infringements on personal privacy have devastating implications for our right to privacy.

Internet privacy is primarily concerned with protecting user information.

Internet users may protect their privacy through controlled disclosure of personal information—sex, age, physical address, email address, etc. Data snooping (an electronic version of eavesdropping) is the process of legally or illegally using technology to gain access to personal information about you. Government uses cybersurveillance to monitor criminals or people they want to track. Organizations implement workplace surveillance to ensure that there is no misuse of time and computing resources. Crackers spy on your personal data and use it for malicious intent. For example, an identity theft would snoop to get credit card number and bank account and use them for their own gain.

Fulfilling customer privacy requirements is often difficult. A number of technologies have been developed in order to achieve information privacy goals. Some of these privacy-enhancing technologies (PET) are described as follows:

- *Virtual private networks* (VPN) are extranets established by business partners. As only partners have access, they promise to be confidential and have integrity.
- *Transport Layer Security* (TLS), based on an appropriate global trust structure, could also improve confidentiality and integrity.
- *DNS Security Extensions* (DNSSEC) make use of public key cryptography to sign resource records in order to guarantee origin authenticity and integrity of delivered information.

Software for computer surveillance includes spyware and adware, which reside on your computer and work like electronic spies.

5.4.2 Security

Although the Internet brings immeasurable opportunities, it also brings new risks. Because of its fast, cheap, and anonymous character, the Internet has become a place for various attacks and criminal activities. By nature, cyberspace or the Internet is difficult to secure. Intruders exploit the vulnerabilities to steal information and money and perpetrate crimes. The crimes include child pornography, banking and financial fraud, and intellectual property violations. They may also include accessing government and defense confidential information, tampering with commercially sensitive data, and targeting supply chains.

Cybersecurity is the process of protecting computer networks from cyber attacks or unintended unauthorized access.

Internet security is not different from other forms of network security. Cyberattacks are threatening the operation of businesses, banks, companies, and government networks. They vary from illegal crime of individual citizen (hacking) to actions of groups (terrorists). The following are typical examples of cyberattacks or threats:

- *Malware*: This is a malicious software or code that includes traditional computer viruses, computer worms, and Trojan horse programs. Malware can infiltrate your network through the Internet, downloads, attachments, email, social media, and other platforms. Spyware is a type of malware that collects information without the victim's knowledge.
- *Phishing*: Criminals trick victims into handing over their personal information such as online passwords, social security number, and credit card numbers.
- *Denial-of-service attacks*: These are designed to make a network resource unavailable to its intended users. These can prevent the user from accessing email, websites, online accounts, or other services.
- *Social engineering attacks*: A cybercriminal attempts to trick users to disclose sensitive information. A social engineer aims to convince a user through impersonation to disclose secrets such as passwords, card numbers, or social security number.
- *Man-in-the-middle attack*: This is a cyberattack where a malicious attacker secretly inserts him/herself into a conversation between two parties who believe they are directly communicating with each other. A common example of man-in-the-middle attacks is eavesdropping. The goal of such an attack is to steal personal information.

Cybersecurity involves reducing the risk of cyberattacks. It involves the collection of tools, policies, guidelines, risk management approaches, and best practices that can be used to protect the cyber environment and mitigate cyberattacks. Cybercrime prevention is a multifaceted issue. Cyber risks should be managed proactively by the management. The best way to protect your network against security breaches is using a firewall, which can protect your network from outsiders.

Internet security objectives have been availability, authentication, confidentiality, nonrepudiation, and integrity:

- *Availability*: This refers to availability of information and ensuring that authorized parties can access the information when needed. Attacks targeting availability of service generally lead to denial of service.
- *Authenticity*: This ensures that the identity of an individual user or system is the identity claimed. This usually involves using username and password to validate the identity of the user. It may also take the form of what you

have such as a driver's license, an RSA token, or a smart card.

- *Integrity*: Data integrity means information is authentic and complete. This assures that data, devices, and processes are free from tampering. Data should be free from injection, deletion, or corruption. When integrity is targeted, nonrepudiation is also affected.

Internet safety refers to how to be safe, confident, and responsible when using online technologies.

- *Confidentiality*: Confidentiality ensures that measures are taken to prevent sensitive information from reaching the wrong people. Data secrecy is important especially for privacy-sensitive data such as user personal information and meter readings.
- *Nonrepudiation*: This is an assurance of the responsibility to an action. The source should not be able to deny having sent a message, while the destination should not deny having received it. This security objective is essential for accountability and liability.

The Internet Protocol Security (IPSec) provides protection to applications that use UDP or TCP. It was developed by the Internet Engineering Task Force (IETF). The IPSec standard is a set of cryptographic protocols that provide secure data exchange, support authentication-level peer networks, and provide data authentication, integrity, and confidentiality (encryption). The original TCP/IP did not provide these security measures. Because IPSec is integrated at the Internet layer (layer 3), it provides security for almost all protocols in the TCP/IP suite.

Cybersecurity policy lags technological innovation. Information technology changes rapidly, with security technology and practices evolving even faster to keep pace with changing threats. Because cybersecurity is not well-understood by non-experts, the economics are hard to demonstrate, and effectiveness is difficult to measure. Minimizing our cybersecurity risks requires commitment on both technical and political fronts.

5.4.3 Safety

Safety is fundamentally important for everyone, whether online or offline, and is everyone's responsibility. Making the Internet safe for children in particular has become a major technological challenge and a public policy issue.

Risks to Internet use include Internet abuse, cyberbullying, privacy violations, and unwanted solicitation. However, claims that cyberspace is dominated by inappropriate material such as pornography, inflammatory, and racist writings are exaggerated. Education on cybersafety may prevent the downside to Internet use. Schools are basically responsible for ensuring Internet safety of young Internet users. Parents can control and lower degrees of unsafe online behavior of their children.

Guidelines for online safety typically include three basic elements: avoiding disclosure of personal information to strangers, creating standards for Internet access, and establishing open communication between children and adults to discuss both positive and negative cyber-experiences. The behaviors related to sharing the sensitive data (such as name and last name, personal pictures, mobile phone numbers, email addresses) on the Internet determine a range of risky behaviors in the cyber world.

Curriculum development is another approach to cope with Internet safety issues. A wide variety of curriculum materials is now available for parents, teachers, and children of different age levels. Children should be informed and taught concrete Internet safety skills. The advantages that children have using the Internet greatly outweigh the risks involved.

5.5 Future Internet

Just as the Internet revolutionized how people communicate and share information, the ongoing development in speed, bandwidth, and functionality will continue to cause fundamental changes. Two leading efforts to define the future of the next-generation Internet are the next-generation Internet Protocol (IPv6) and Internet2, which are discussed in this section.

5.5.1 IPv6

Most of today's Internet uses Internet Protocol Version 4 (IPv4), which is now very old. Due to the phenomenal growth of the Internet, the rapid increase in palmtop computers, and the profusion of smart phones and PDAs, the demand for IP addresses has outnumbered the limited supply provided by IPv4. Four of the five regional internet registries (RIRs) have depleted their IPv4 allocations and began operating under final IPv4 address depletion policies. IPv4 is designed to have a 32-bit address field, limiting the total number of unique IPv4 addresses to approximately 4.3 billion (to be exact, 2^{32} addresses, which is 4,394,967,296). While this seems like a large number, some of these addressed are unused and inefficiently assigned.

In response to this depletion of IPv4 addresses, the Internet Engineering Task Force (IETF) approved Internet Protocol Version 6 (IPv6) in 1997. IPv4 will be replaced by IPv6, which is sometimes called the next-generation Internet Protocol (or IPng). The IETF specification for IPv6 is RFC 2460. IPv6 adds many improvements and fixes a number of problems in IPv4, such as the limited number of available IPv4 addresses.

IPv6 is the next generation protocol designed to solve IPv4 address depletion problem.

People are looking toward IPv6, with its 2^{128} possible addresses, as the solution. IPv6 is designed to support a global public Internet and has an address space of 128 bits, allowing for more than 340 undecillion (10^36) unique IPv6 addresses (340,282,366,920,938,463,463,374,607,431,768,211,456 to be exact!). All future Internet growth will be over IPv6.

The main features of IPv6 include:

- A larger address space
- Optimization of multicast addressing and routing anycast appearance
- Use of extension headers and a flow labeling mechanism
- Intrinsic safety in the core protocol
- Quality of service and service classes
- More efficient backbone routing
- Renumbering and multihoming, which facilitates the change of service provider
- Supports a network node to configure its address by itself
- Supports mobile applications better than IPv4
- Support for IP security (IPsec)

The IPv6 header consists of 40 bytes as shown in Fig. 5.5. The fields are explained as follows:

- *Version* (4 bits): This represents the IP version number, which is 6 (bit sequence 0110).
- *Traffic class* (8 bits): This indicates priority values, which are subdivided into ranges: traffic where the source provides congestion control and non-congestion control traffic.
- *Flow label* (20 bits): When set to a non-zero value, it hints the routers that these packets should stay on the same path. A label is used to maintain the sequential flow of the packets belonging to a communication. It is designed for streaming/real-time media.
- *Payload length* (16 bits): This specifies the size of the remaining part of the packet following the header (payload) in octets.
- *Next header* (8 bits): Specifies the type of the next header immediately following this.
- *Hop limit* (8 bits): Replaces the time to live field of IPv4. This value is decremented by one at each intermediate node visited by the packet. The packet is discarded when the counter reaches 0. This field is used to stop packet from looping in the network infinitely.
- *Source address* (128 bits): The IPv6 address of the sending node.
- *Destination address* (128 bits): The IPv6 address of the destination node(s).

For more details, please see RFC 2460–IPv6 Specification.

IPv6 addresses are 128 bits in length and are made up of hexadecimal characters. They are expressed as a series of 8 4-character hexadecimal numbers, which represent 16 bits each (for a total of 128 bits). For examples the IPv6 addresses are:

2001:E10:C41:49:00C5:281F:44A8:80BD
1201:DB8:85A3:0:0:8A2E:370:2134
2601:2C3:8580:7A10:25B5:5F03:0000:441C

We notice that these addresses are large and not easy to work with. One or more consecutive groups of zero value may be replaced with a single empty group using two consecutive colons (::). Thus, the second address can be further simplified:

2001:DB8:85A3::8A2E:370:2134

IPv6 addresses are classified into three types as unicast addressing, anycast addressing, and multicast addressing. A unicast (one-to-one) address identifies a single network interface. An anycast (one-to-nearest) address is assigned to a group of interfaces, usually belonging to different nodes. Almost any unicast address can be employed as an anycast address. A multicast (one-to-many) address is also used by multiple hosts. IPv6 does not implement broadcast addressing; it is replaced by multicast addressing. DNS is still used in IPv6 (Fig. 5.6).

IPv6 has experienced a slow deployment and is still perceived as experimental by some. Many organizations and Internet stakeholders are not enthusiastic in implementing IPv6 because they see no compelling need for it. An important step has been made by the US government regarding the transition to IPv6 for all governmental websites in the USA by the end of 2012. Companies like Comcast, AT&T, and Time Warner Cable made the transition to IPv6. Facebook is an example of an organization which ran out of RFC 1918 addresses and has implemented an IPv6 solution.

IPv6 adoption has been slow because IPv6 is not generally compatible with IPv4 networks. As a result, a number of transition technologies use tunneling to facilitate cross network compatibility. To perform transition from IPv4 to IPv6, tunneling is used as a transition strategy, as shown in Fig. 5.7. Tunneling is used to connect two IPv6 nodes using IPv4 network. It encapsulates packets of one protocol onto another. It is expected that IPv4 and IPv6 will coexist for a while and that IPv6 tunnels play a key role in the transient phase.

5.5.2 Internet2

The current, original Internet (now referred to as Internet1) was not designed to handle applications such as online education, e-commerce, and interactive multimedia. Applications such as videoconferencing, uncompressed HDTV, gigabit data set sharing, weather forecasting, satellite imaging, telemedicine, and remote vision require high bandwidth and real-time response. The deficiencies in Internet1 (such as congestion and reli-

Offsets	Octet	0								1								2								3							
Octet	Bit	0	1	2	3	4	5	6	7	8	9	10	11	12	13	14	15	16	17	18	19	20	21	22	23	24	25	26	27	28	29	30	31
0	0	Version				Traffic Class								Flow Label																			
4	32	Payload Length																Next Header								Hop Limit							
8	64	Source Address																															
12	96																																
16	128																																
20	160																																
24	192	Destination Address																															
28	224																																
32	256																																
36	288																																

Fixed header format

Fig. 5.6 IPv6 header format

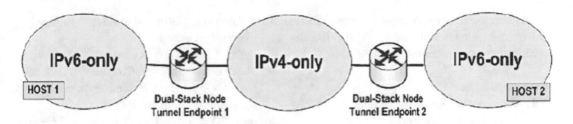

Fig. 5.7 Tunneling

ability, limited IP addresses, limited bandwidth) have frustrated millions of users. As a result, Internet2 was created in 1996 by the research and education community to serve its advanced networking needs and meet its bandwidth-intensive requirements. The idea of Internet2 was initiated by 34 US universities collaborating with government and industry. Internet2 now "serves more than 90,000 community anchor institutions, 305 U.S. universities, 70 government agencies, 42 regional and state education networks, 84 leading corporations working with our community and more than 65 national research and education networking partners representing more than 100 countries."

network will not solve all the problems of Internet1. Internet2 is still evolving. It has partners in Europe, Africa, Asia, and Middle East with the goal of connecting research communities around the globe. For more information on Internet2, see http://www.internet2.edu.

Summary
1. The Internet is a global "network of networks" of autonomous networks that support communication and collaboration among users. It is based on TCP/IP protocol suite.
2. A protocol is a set of rules and procedures followed by two or more computers in communicating.

The **Internet2** is designed to establish a next-generation Internet which will enable scholars to collaborate with colleagues around the globe.

Abilene is one of Internet2's backbones, using optical transport technology and high-performance routers. Fiber optics will be a major role in developing infrastructure for Internet2. It provides high-bandwidth data highways for connecting regional and global communities. Internet2 consortium intends to upgrade Abilene high-speed network backbone and implement IPv6 (Internet Protocol Version 6) while continuing to support IPv4.

Major beneficiaries of Internet2 will be research libraries, digital videoconferencing, multimedia presentation, medical imaging, massive electronic submission, distance education, and teleconsultation. The benefits of Internet2 are endless and its future holds much promise.

While Internet2 as an organization is elitist, its goals are worthy and achievable. Internet2 as a

3. IP operates at the network layer and is connectionless.
4. TCP and UDP are two protocols that operate on the transport layer.
5. The most important Internet services include emailing, file transfer, remote login, and web browsing.
6. Important Internet applications include voice over IP, social networking, education, e-commerce, and government.
7. The issues privacy, security, and safety are important in various uses of the Internet. Privacy deals with how to handle privacy violations. Security deals with how information can be protected. Safety refers to how to be safe online.
8. Future Internet is in the form of IPv6 and Internet2. IPv6 was introduced to solved IP address depletion problem of IPv4. Internet2

was created to serve Internet's advanced networking needs and meet its bandwidth-
intensive requirements.

Review Questions

5.1. Which of the following switching methods
is used by the Internet?
(a) Packet switching   (b) Circuit switching   (c) Cable switching   (d) Transmission switching

5.2. Internet supports message exchange
between users through a mechanism called:
(a) Gateway   (b) Interfaces   (c) Layers   (d) Protocols

5.3. Which protocol provides a connectionless
service?
(a) IP   (b) TCP   (c) UDP

5.4. APNIC is responsible for allocating IP
addresses for:
(a) Africa   (b) America   (c) Asia and the
Pacific region   (d) Latin America and
the Caribbean region   (e) Europe,
Middle East, and Central Asia

5.5. Which of the following is not true?
(a) The Internet is a specific piece of hardware and software.
(b) The Internet is not the same
everywhere.
(c) The Internet is not restricted to educational, noncommercial uses.
(d) The Internet is a medium for communicating with others.
(e) The Internet is flexible in cost.

5.6. Only classes A, B, and C of IP addresses
are commonly used.
(a) True   (b) False

5.7. The IP address 152.116.24.100 belongs to:
(a) Class A   (b) Class B   (c) Class C   (d)
Class D   (e) Class E

5.8. The most popular Internet service for commercial purposes is:
(a) Email   (b) Telnet   (c) FTP   (d)
WWW

5.9. IPv6 does not support the following
addressing.
(a) Unicast addressing   (b) Anycast
addressing   (c) Multicast addressing   (d) Broadcast addressing

5.10. Internet2 is designed to solve the following
problems except:
(a) Limited IP addresses   (b) Limited
bandwidth   (c) Congestion and reliability   (d) Connecting regional and
global communities

Answer: 5.1 a, 5.2 d, 5.3 c, 5.4 c, 5.5 a, 5.6 a, 5.7
b, 5.8 d, 5.9 d, 5.10 d

Problems

5.1. Write about two individuals who played
key roles in the development of the
Internet.

5.2. Describe the responsibilities of the four
layers in the Internet protocol stack.

5.3. Why is Internet regarded as a "network of
networks"?

5.4. What is the major difference between TCP
and UDP?

5.5. Explain fields IHL, DSCP, ECN, and TTL
of an IP header format.

5.6. Convert the following IP addresses in
binary to decimal.
(a) 10000010   01110101   00110011   1
1110001
(b) 11111110   01010101   00000000   1
1001100

5.7. What is the class of each of the following
IP addresses?
(a) 244.35.36.1
(b) 87.100.5.105
(c) 232.46.55.111
(d) 190.205.0.10

5.8. Specify which class each of the following
IP addresses belongs.
(a) 10000010   01110101   00110011   1
1110001
(b) 11100010   01110000   00100010   1
0000001
(c) 01000011   00001010   10011001   1
0001110
(d) 11010011   00001000   11000011   0
0010100

5.9. Discuss three Internet services.

5.10. Discuss the following protocols: RTP,
RTCP, and ARP.

5.11. What is VOIP? What are its major
benefits?

5.12. What is a VOIP gateway responsible for?

5.13. What are the major services online education must provide?

5.14. What is Massive Online Open Course (MOOC)?

5.15. What is e-commerce? What advantages does it have over traditional commerce?

5.16. Discuss online banking.

5.17. In what ways has the Internet impacted modern political life?

5.18. Explain the concept of open government.

5.19. Discuss IPSec.

5.20. Discuss three cyberattacks or threats.

5.21. What are the built-in advantages of IPv6 over IPv4?

5.22. Discuss the following fields of IPv6 header format: flow label, payload length, and next header.

5.23. Discuss the three types of IPv6 addresses.

5.24. Discuss IPv6.

5.25. Simplify the following IPv6 addresses:
(a) 0000:0000:0000:0000:349A:2651:000 0:CB12
(b) 6127:0000:0000:0000:9010:0024:FA1 1:DE28
(c) 0023:0000:0000:72FC:0000:0000:390 0:AD41

5.26. What is Internet2? Who will benefit from it?

Intranets and Extranets

6

Major power and telephone grids have long been controlled by computer networks, but now similar systems are embedded in such mundane objects as electric meters, alarm clocks, home refrigerators and thermostats, video cameras, bathroom scales, and Christmas-tree lights – all of which are, or soon will be, accessible remotely.

Charles C. Mann

Abstract

This chapter cover intranets and extranets as two of the growing applications of the Internet. An intranet is a closed or restricted network, only used by those in the company that owns it. It uses the same kinds of links, routers, and protocols as the public Internet, but it is only accessible within the company. An extranet is an extension of an intranet. It is the part of the intranet that outsiders can use. Through extranets, a company can make only a portion of its intranet accessible to external parties. Both intranets and extranets are applied in sales, database access, human resources, customer service, health service, financial services, and manufacturing.

Keywords

Intranets · Extranets · Internet

6.1 Introduction

The Internet is flourishing and Internet connectivity is growing at an exponential rate. A fundamental concept of *Intranet*, the so-called second wave, was introduced. Medium- and large-sized companies use internet features internally (called intranets). The earliest intranets came into being during the 1990s. The term "intranet" was coined toward the end of 1995. Since 1997, more than nine out of ten major companies have implemented their intranet. The new era of the *extranet* or the *third wave* of the universal Internet came into being.

An intranet is simply a closed or restricted network, only used by those in the company that owns it. It uses the same kinds of links, routers, and protocols as the public Internet, but it is only accessible within the company. Campus and corporate networks are good examples. An intranet should be closed to the general public and require a login for authorized users.

Extranets come into the picture because it is prudent to extend this efficiency to the outside world. Through extranets, a company can make only a portion of its intranet accessible to external parties. For example, a corporate intranet may dedicate a number of servers for vendors and other outsiders. A campus extranet would allow parents of college students to log in and pay tuition and look at other information. But it would not give outsiders access to the actual intranet. Extranets are still closed to the general public and require some type of authentication. The Internet, intranet, and extranet will revolutionize how business people, lawyers, educators, etc. conduct their businesses and communicate.

Table 6.1 Comparing the characteristics of the Internet, intranet, and extranet

	Internet	Intranet	Extranet
Access	Public	Public	Semi-private
Users	Everyone	Employees of specific organization	Employees of related organizations
Information	Fragmented	Proprietary	Information only for trusted partners

Source: Askelson, 1998

A comparison of the characteristics of the Internet, intranet, and extranet is given in Table 6.1.

Intranets support intrabusiness activities, while extranets support business-to-business (B2B) e-commerce. An intranet will target managers, sales representatives, executives, and other employees within an organization. Extranets instead target privileged customers, suppliers, contractors, partners, and other possible end users. Intranets are most often found in banks, IT companies, manufacturers, large retail companies, and service companies such as hotels and travel agents. Extranets are often found in IT services, computer hardware companies, financial services, manufacturers, real estate, and professional services.

This chapter presents an introduction to intranets and extranets as two of the growing applications of the Internet.

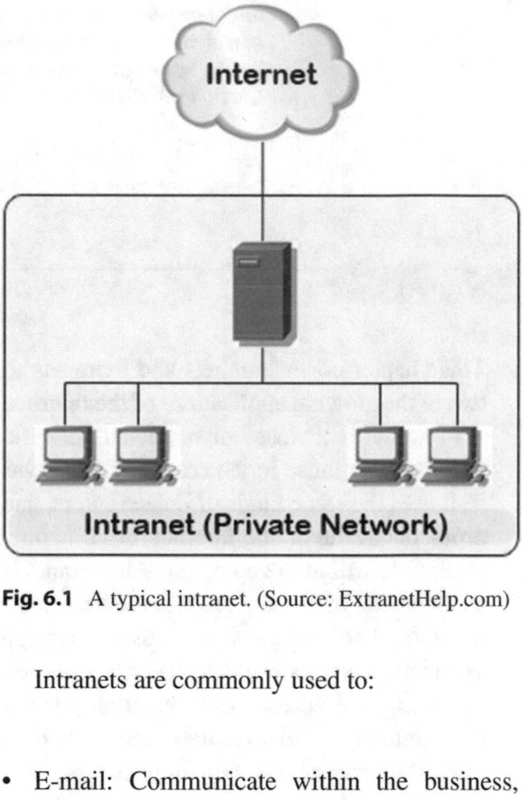

Fig. 6.1 A typical intranet. (Source: ExtranetHelp.com)

6.2   Intranets

The excitement created by the Internet (a public network) has been transferred to an application called intranets. These are private company network, whose users are often the company's employees. "Intranet" means "internal" or "within," so an intranet is an internal or private network that can only be accessed within the confines of a company, university, or organization.

An intranet is connected to the Internet through one or more firewalls. Components of an intranet include servers, networks, and web browsers. An intranet uses HTML to create documents and TCP/IP to transmit data across the network. A typical intranet is shown in Fig. 6.1.

Intranets are commonly used to:

- E-mail: Communicate within the business, share information, and provide feedback.
- Directory: Intranets offer a centralized location for hosting vital business information.
- File sharing: Enable employees to do their jobs in an increasingly globalized, digital workplace by connecting employees across organizations and helping them to thrive. Provides job posting and opportunities.
- Collaboration: Enable collaboration and provide tools that can support employees to work together irrespective of their department or location.

An **intranet** is a web-based network within an organization where employees can create, communicate, work, collaborate, and develop the organization culture.

- Searches: Provide search engines and directories for companies handling massive quantities of text documents, e.g., insurance and banking companies.
- Management: Integrate with electronic commerce Internet-based sales and process orders, e.g., Amazon.com.

The need for intranets has been driven by the push of technology standards and the pull of organizational need to communicate across geographic barriers. Described as "mini-Internets," intranets are transforming the way government and organizations function. They provide a low-cost alternative to distributing paper copies of documents.

Advantages of intranets include the following: (1) It offers workplace productivity; (2) it allows communication within the company; (3) it enables companies to share out information to employees according to their need or requirements; (4) it saves time and money and increases work efficiency since there is no need to maintain physical documents such as procedure manual, requisition forms, and Internet phone list; (5) it allows users to access multiple applications and tools from one place, often using a single login; (6) it is designed to save companies money through network efficiency, productivity, and mobility; (7) it promotes common corporate culture; and (8) it publishes important company's announcements.

Disadvantages of intranets include the following: (1) management has no control over specific information, (2) security is a major issue, and (3) the cost of implementing and maintaining intranets is very high.

6.3   Extranets

An extranet is an extension of an intranet. It is the part of the intranet that outsiders can use. It is a public network that is controlled and shared on a limited basis without granting access to the company's entire network. The term "extranet" refers to an extranet providing various levels of accessibility to outsiders. As typically shown in Fig. 6.2, an extranet is basically an intranet that provides password-protected partial access to selected parts of the intranet to authorized outsiders or external users such customers, partners, and contractors.

An **extranet** works like an **intranet**, but also provides controlled access to authorized customers, vendors, partners, or others outside the organization.

As an extended Internet, extranet is a private business network of several cooperating organizations located outside the corporate firewall. An extranet depends on the existing Internet infrastructure such as servers, e-mail clients, and web browsers. Extranet users can benefit from SMTP, HTTP, and MTML like Internet users, but can only have access to the extranet if they are given username and password. Thus, extranets are often constructed using the Internet with security features that restrict access to authorized individuals. A firewall is used to limit access to a company's intranet. The extranet is regarded as the key technology, which can enable development of the third wave of electronic commerce sites.

Extranets are commonly used to:

- Training: Provide online training programs or other educational material to employees and resellers.
- Sharing: Share product catalogs and newsgroups and post jobs.
- Backbone network: Provide an economical backbone for enterprise networks.
- Maintenance: Maintain closer contact with employees, customers, and suppliers. Maintain and enhance existing customer relationship.

Several extranets have been constructed. Examples include Bank of America (in San Francisco), Cigna (in Philadelphia), Kinko's (in Ventura), The Link (in Los Angeles), Nasdaq (in New York), First Union Bank (in Charlotte), Kroger (in Cincinnati), Lucent Technologies (in

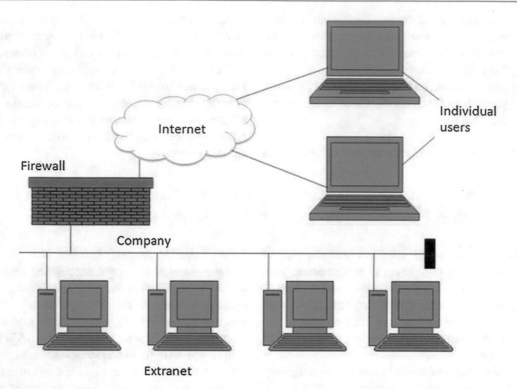

Fig. 6.2 A typical extranet. (Source: "Extranet overview," https://www.tutorialspoint.com/internet_technologies/extranet_overview.htm)

Murray Hill), and Microsoft (in Redmond). Other companies with extranets include FedEx, Shell, Cisco, McDonnell Douglas, and Courtyard Marriott. Many extranets serve the purpose of showing demos and examples. Surveys, feedback forms, FAQs, statistical data, and other forms of improving an organization's infrastructure are popular among extranets.

Different types of extranets may be needed for different applications. For example, an extranet may be needed for high-volume trans-actions or videoconferencing. A retailer extranet offers an extranet to supply authorized chain partners to facilitate processes such as invoice payment and reverse logistics. An employee extranet allows employees to submit any con-cerns or recommendations they have without their manager seeing it.

Closely associated with intranets and extranets are portals. A portal is a door leading to informa-tion areas in an intranet and the Internet. The por-tal is an information gateway that adds to the unity and consistency that intranets provide an

organization. There are different types of portals. Intranet web portals provide links to all internal content providers. Information portals present a uniformed look for access through system login. Knowledge portals provide information catalog and a knowledge repository.

Advantages of extranets include (1) exchang-ing large volumes of data, (2) sharing product catalogs exclusively with trade partners, (3) col-laborating with other companies and developing training programs with them, (4) providing ser-vices to other companies, (5) enabling e-business or business-to-business e-commerce, and (6) pro-moting more efficient customer and supplier relationships.

Disadvantages of extranets include the follow-ing: (1) extranets can be expensive to implement since it involves hardware, software, and employee training costs; (2) security of extranets is a major concern when hosting valuable or pro-prietary information as well as commercial trans-actions such as electronic funds transfer. Security, at the network and host levels, becomes critical within an extranet. This is why managing access control, which involves defining user access privileges, managing passwords, user IDs, and authentication schemes, is at the heart of every extranet.

The benefits of extranets (communications, collaboration, and coordination) far outweigh the initial deployment headaches. A typical application of extranets would be in the data center where all of the servers are located. This can be easily implemented by creating a virtual private network (VPN) for each department. Many organizations with mature intranets are now connected to the Internet and operate the two networks as if the components were a single network, without borders. The advent of extranets is revolutionizing supply chain management..

6.4   Key Technologies

Intranets and extranets are constructed with the same three basic components: computer network or networks, client software, and server software. They are related as shown in Fig. 6.3. They are no longer luxuries. They provide information in a cost-effective way. Their applica-

tions are scalable, i.e., they can start small and grow.

Three key developments are necessary for implementing intranets and extranets. First, there must be the acceptance of widely used standards for communication across networks. Second, there must be the development of compression and decompression techniques to transmit large amounts of data over networks. The image compression and decompression technology in the form of Joint Photographic Experts Group (JPEG) compression are a key in imaging applications. Third, secure public communication is a major concern when hosting valuable or proprietary information (extranet), a vital asset. It is not necessary if information is restricted to a private network (intranet). Security takes on a new dimension with extranets since companies must protect internal systems from unwanted visitors.

These core requirements of extranets make them to be open, portable, and interoperable with different industry standards across multiple platforms.

Both intranets and extranets use web technology. The five key elements that make web technology possible and cost-effective include as

Fig. 6.3 Relationship between the Internet, Intranet, and Extranet. (Source: "Difference between Internet and Extranet," November 2020, https://www.geeksforgeeks.org/difference-between-internet-and-extranet/)

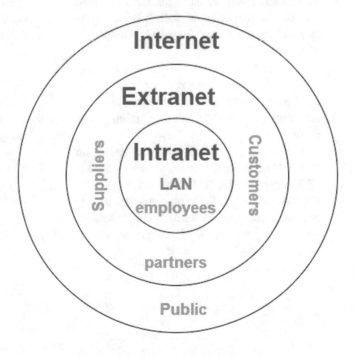

follows: (1) The communications protocol of the Internet (HTTP—HyperText Transport Protocol), (2) the almost universal acceptance of the IP protocol as a preferred networking protocol standard, (3) standard browsers such Internet Explorer and Google Chrome, (4) powerful web technology tools, and (5) platform-independent nature of the technologies. The networks are protected from outside access by firewall, which often serves as a link between the company and the public Internet.

Both intranets and extranets are applied in sales, database access, human resources, customer service, health service, financial services, and manufacturing. Numerous companies that have successfully implemented them include AT&T, IBM, HCA Healthcare, FedEx, Ford Motor, Home Depot, and US West.

Summary

1. An intranet is a corporate local area network that uses World Wide Web technologies and Internet communication protocols.
2. An intranet is used to share information, collaboration, and management.
3. An extranet, or extended Internet, is a private business network of an organization located outside the corporate firewall. It is an expansion of intranet that can be used by company's partners.
4. The extranet is regarded as the key technology, which can enable development of the third wave of electronic commerce sites.
5. An extranet is used for sharing, training, and maintenance.
6. Both intranet and extranet are based on TCP/IP and related protocols such as SMTP, FTP, secure HTTP, XML, etc.
7. Key developments for implementing intranets and extranets are standards for communications, compression and decompression techniques, and security in public communications.

Review Questions

6.1. An intranet is used for the following except:
(a) Management (b) File sharing (c) Collaboration (d) Advertisement

6.2. An intranet can be accessed from outside a company.
(a) True (b) False

6.3. Advantages of intranets include the following except:
(a) Productivity (b) Efficiency (c) Marketability (d) Sharing information

6.4. Disadvantages of intranets include the following except:
(a) Security (b) Implementation cost (c) Lack of control over information (d) Sharing directory

6.5. An extranet is basically an intranet that is partially accessible to authorized outsiders.
(a) True (b) False

6.6. Extranets are used for the following except:
(a) Education (b) Advertisement (c) Conference meeting (d) Backbone network (e) Maintenance

6.7. The benefits of extranets include the following except:
(a) Conferencing (b) Communication (c) Collaboration (d) Coordination

6.8. Advantages of extranets include the following except:
(a) Exchanging information (b) Collaboration (c) Being public (d) Enabling e-commerce

6.9. Intranets and extranets are constructed using the Internet with security features.
(a) True (b) False

6.10. Virtual private network (VPN) is:
(a) An intranet (b) An extranet (c) Both

Answer: 6.1 d, 6.2 b, 6.3 c, 6.4 d, 6.5 a, 6.6 b, 6.7 a, 6.8 c, 6.9 a, 6.10 c

Problems

6.1. What is the meaning of intranet? Extranet?

6.2. What are intranet and extranet made of?

6.3. How is intranet used?

6.4. What are examples of intranets?

6.5. What are the advantages and disadvantages of intranets?

6.6. What are the similarities between intranet and extranet?

6.7. Mention five specific extranets.

6.8. How is extranet used?

6.9. Why do we have different extranets?

6.10. What is the difference between Internet and intranet?

6.11. List the advantages and disadvantages of an extranet.

6.12. Why is security a major issue in extranets?

6.13. Some consider extranet as a private network, while others see it as a public network. Explain.

6.14. What are intranet and extranet made of?

6.15. What is access control in extranet? Why is it important?

6.16. Why are there different types of extranets?

6.17. What is a portal?

6.18. What are the three key developments required for implementing intranet and extranets?

6.19. Mention the five key elements that make web technology possible for intranets and extranets.

6.20. List five major applications of intranets and extranets.

6.21. Mention five companies that have successfully implemented intranets and extranets.

Virtual Private Networks

7

The most compelling reason for most people to buy a computer for the home will be to link it to a nationwide communications network. We're just in the beginning stages of what will be a truly remarkable breakthrough for most people—as remarkable as the telephone.

Steve Jobs

Abstract

This chapter discusses virtual private network (VPN). A VPN combines the advantages of both public and private networks. It is designed to create an encrypted tunnel between you and your provider. It is a mechanism for simulating a private network over a public one. The connectivity between the Internet and VPN sites is through a security device such as a firewall. The application scope of VPN is increasing day by day as the organizations are creating private networks through public Internet instead of leased line. VPN services have been offered in various forms over the years. Such services include ecommerce, social networking, mobile networks.

Keywords

Virtual private network · Internet · Firewall · VPN services

7.1 Introduction

Companies have realized that remote access is a strategic weapon in the increasingly competitive global economy. To provide remote access and ensure secure and reliable communication, most enterprises used private leased line to connect remote users or offices. A leased line (from phone companies) connects two endpoints on a network to make sure that the connection is secure and private. The main problem with the traditional leased line is that it takes a lot amount of time to deploy, and it is expensive since the subscribers must pay for the bandwidth all the time, including when the line is idle. As the use of the Internet has grown, businesses have found that the Internet is an inexpensive way for mobile users and fixed sites to connect to the corporate network. Clearly it is cheaper to connect over the Internet to a private network than to pay for a leased line to achieve the same goal. Modern organizations use alternate secure and private communication networks like virtual private network (VPN). A VPN combines the advantages of both public and private networks. It is a mechanism for simulating a private network over a public one.

As shown in Fig. 7.1, a virtual private network (VPN) is a network, built on top of existing public or non-private network through the use of a tunneling protocol, to provide a secure communications between two endpoints. First, a VPN is a "network" in that it provides interconnectivity to various components of the VPN. It is a combination of hardware and software that allows multi-

© The Author(s), under exclusive license to Springer Nature Switzerland AG 2022
M. N. O. Sadiku, C. M. Akujuobi, *Fundamentals of Computer Networks*,
https://doi.org/10.1007/978-3-031-09417-0_7

Fig. 7.1 A typical
example of VPN.
(Source: Denise
Grayson et al., 2009, pp.
146–163)

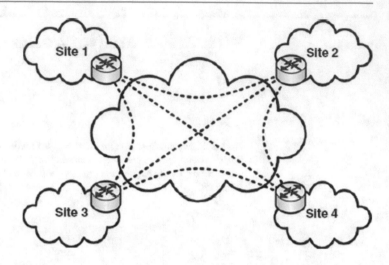

ple devices to communicate with each other. Second, it is "virtual" because it is a connection that has the appearance of a dedicated line but occurs over a shared public network. The VPN uses virtual connections, i.e., temporary connections that have no physical presence. The VPN creates the illusion of a completely private network. Third, it is "private," meaning that it is only used by the enterprise and is inaccessible to others. It is also private in the sense that the routing and addressing scheme within the network is totally independent of the routing and addressing plans of all other networks. It guarantees privacy within the organization and provides limited access to only authorized users.

A VPN is designed to create an encrypted tunnel between you and your provider. It provides a secure, encrypted tunnel to transmit the data between the remote user and the company network. It consists of special hardware augmented with software designed to provide VPN functions.

and VPN sites is a through a security device such as a firewall.

VPN uses tunneling and encryption to do the following:

- It provides privacy by hiding your Internet activity from your Internet service provider (ISP). The ISP cannot see your data because it is encrypted.
- It allows one to avoid censorship by government, ISP, company, etc.
- It protects one against hackers when using a public Wi-Fi. VPN protects your data using encryption.

A VPN can be used to interconnect two similar networks over a dissimilar middle network. It can also be used to securely connect geographically separated offices of an organization. VPN technology can be used to supplement or replace frame relay or leased lines. Most enterprises supplement their WANs with VPNs.

A **virtual private network** is a private network running over a public network like the Internet.

A VPN is designed to provide wide area connectivity to a company located in multiple sites. It secures one's computer's Internet connection and ensures that all of the information one is sending and receiving is encrypted and secured from prying eyes. The connectivity between the Internet

This chapter covers the main characteristics of VPN, different types of VPNs, various applications of VPNs, and the benefits and challenges of VPNs. The chapter concludes with a summary of the items covered in the chapter.

7.2   VPN Characteristics

VPN characteristics are crucial in selecting an optimal solution for a given application. VPNs have the following characteristics.

Deployment: VPNs can be either remote access VPN or site-to-site VPN. Remote access VPNs enable employers, business partners, suppliers, and customers to connect to enterprise network from home or remote location. Site-to-site VPNs are used to connect branch offices, and they generally supplement or replace leased lines or frame relay. Such VPNs link two dispersed networks as they might be linked by a leased line or a WAN circuit.

Tunneling: Tunneling is a characteristic of all VPNs. If the data being tunneled gets encrypted, you have a VPN, as illustrated in Fig. 7.2. Some VPN protocols bear the term "tunneling," such as Microsoft's Point-to-Point Tunneling (Protocol PPTP), to be discussed shortly. Although tunneling can happen at virtually any layer, the two common types of tunneling are at Layer 2 and Layer 3. Layer 3 provides the underlying structure to set a VPN connection at the application layer. Modern mobile devices are powered with Layer 3 tunneling protocol, such as IPSec, to be discussed shortly.

Security: This is the main reason why organizations have used VPNs for years. Security is one of the key requirements for business success. It ensures that sensitive data can be transported over the network. The justification for using it instead of a private network is due to cost and feasibility. It may be too expensive to have a private network. One can simply think of a VPN as a secure tunnel between your PC and destinations you visit on the

Internet. Once you are connected to the VPN, it is difficult for anyone else to spy on your web-browsing activities. Suppose you live in Texas and your VPN provider is based in New York, your traffic will look like you are connecting from New York. The problem with that is that if you search for "pizza delivery," you will get shops that deliver in New York.

Safety: To ensure safety, VPN users must use authentication methods to gain access to the VPN. Standard network requirements like confidentiality, integrity, and authentication are ensured by the VPN. Confidentiality ensures that measures are taken to prevent sensitive information from reaching the wrong people. Data secrecy is important especially for privacy-sensitive data such as user personal information and meter readings. Data integrity means information is authentic and complete. This assures that data, devices, and processes are free from tampering. Data should be free from injection, deletion, or corruption. When integrity is targeted, nonrepudiation is also affected. Authenticity ensures that the identity of an individual user or system is the identity claimed. This usually involves using username and password to validate the identity of the user. It may also take the form of what you have such as a driver's license, an RSA token, or a smart card.

Topology: Traditional VPNs are characterized by a point-to-point topology. One can create a VPN by establishing a virtual point-to-point connection through the use of dedicated connections. This point-to-point topology is a major limitation of traditional VPN because they cannot connect broadcast domains. This limitation is overcome by a variant of VPN known as Virtual Private LAN Service (VPLS).

Fig. 7.2 Tunneling forming a virtual private network connection

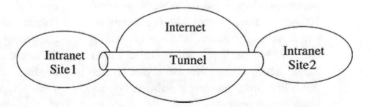

Encryption: For VPN to ensure security, data is encapsulated and encrypted before sending the packets over the Internet. The encryption is done using Microsoft Point-to-Point Encryption (MPPE) protocol. Using VPN will slow down your Internet activity because encrypting and decrypting data takes processing time.

Protocols: Some of the commonly used tunneling protocols in VPNs are:

- *Point-to-Point Tunnelling Protocol* (PPTP): This was developed by Microsoft [RFC 2637]. It has relatively low overhead, making it the fastest among the various VPN tunneling methods. PPTP is one of the remote access tunneling techniques that creates private virtual point-to-point connection. It merely establishes the tunnel but does not provide encryption.
- *The Layer 2 Tunnelling Protocol* (L2TP): This was developed by Cisco and Microsoft, combining features of PPTP and Layer 2 (data link layer) forwarding (L2F) protocol. The L2TP is another highly secure remote access tunneling protocol. It is more effective than PPTP for remote access VPN. Its major advantage is that it can be used on non-IP-based networks such as frame relay and X.25.
- *Internet Protocol Security* (IPSec): This (secure IP) is regarded as the "standard" for VPNs that connect two LANs. It operates at Layer 3 (network layer). It was standardized by the IETF in the late 1990s (as RFCs 1825 through 1829) and is used for achieving secure communications over the Internet, securing the network at the IP level with encryption and authentication. Site-to-site VPN protocol IPSec is more applicable for security-sensitive applications. IPSec is becoming the VPN protocol of choice over PPTP. Many companies use IPSec to create VPNs that run over a public network such as the Internet.

7.3 Types of VPN

There are several types of VPN. These include:

Remote access VPN: The most popular VPN is the remote access VPN. With the workforce becoming increasingly mobile, providing employees remote access to the corporate network is necessary. A remote access VPN allows remote corporate users to have on demand connectivity into their corporate intranets through tunneling. In the past, remote access has been established by dialing the telephone network. For this reason, remote access VPN is also called dial-in VPN. However, a new protocol called L2TP [2] allows one to set up PPP connections over the Internet.

Intranet/extranet VPN: The other common VPN type is intranet/extranet VPN. We recall that in Chap. 6 that an intranet is a private computer network that uses Internet protocol, leveraging existing technologies based on TCP/IP. When the enterprise shares some of its intranet resources with the suppliers, vendors, partners, customers, or other businesses, the expanded network is called extranet. An extranet can be viewed as part of a company's intranet that is extended to users outside the company (e.g., normally over the Internet). An intranet/extranet VPN links the network of a headquarter office to the networks of remote branches and potentially to the networks of suppliers, partners, customers, and other communities of interest. The Automotive Network eXchange (ANX) is a good example of VPN extranet. It is a cooperative, flexible network for automotive industry, patronized by the big three (GM, Ford, and Chrysler).

Cloud VPN: Cloud computing consists of a set of resources and services offered through the Internet or its dedicated network. Hence, "cloud computing" is also called "Internet computing." The word "cloud" is a metaphor for describing web as a space where computing has been preinstalled and exists as a service. Operating systems, applications, storage, data, and processing capacity all exist on the web, ready to be shared among users. Cloud computing provides three basic architectures to take care of the needs of different users: IaaS (Infrastructure as a Service), PaaS (Platform as a Service), and SaaS (Software as a Service). Most of the organizations with VPN would like to reduce the operational cost, implementation cost, and overhead associated with their VPN. The data protection and confidentiality of sensitive information in cloud environment are assured by deploying encryption, which can be achieved by setting up VPN services in cloud.

Unicast VPN: This basically uses Layer 3 multiprotocol label switching (MPLS) as the encapsulating protocol. The ingress provider edge router attaches an MPLS label to a packet it receives from the custom. This label is used in forwarding the packet to the egress provider edge router, where it is decapsulated and passed to the attached customer. This is also called MPLS-VPN (RFC 2547), and it can secure wireless mesh networks, the dual deployment of both technologies. Enterprise networks are increasingly adopting

MPLS-VPN technology to connect geographically disparate locations. However, the any-to-any direct connectivity model of this technology is causing rapid increase in the routing tables in the service provider's routers.

Multicast VPN: The growth of one-to-many applications in corporate networks has led to the demand for VPNs to support multicast applications. In a multicast VPN (or MVPN), customer multicast traffic is encapsulated at an ingress provider edge router. The traffic is delivered to one or more egress provider edge routers.

Other types of VPN include peer-to-peer VPN, Layer 2 VPN, and IPSec VPN. Commercially available VPNs include Private Internet Access VPN (with monthly charge of $6.95), NordVPN ($11.95 per month), VPN Unlimited ($5.99 per month), IPVanish VPN ($5.99 per month), and ExpressVPN ($8.32 per month). Generally, all VPNs make a public network

(like the Internet) work like a private network and encrypt the data.

7.4 Applications

The application scope of VPN is increasing day by day as the organizations are creating private networks through public Internet instead of leased line. VPN enables a corporate network to stretch around the globe and provide corporate information, no matter where it is stored, in seconds. It offers voice over IP, streaming video, instant messaging, and crowd computing. It is all done at Internet service prices.

VPN technologies play a vital role in various areas such as education, businesses, retail, finance, manufacturing, government, and military organizations. For example, in modern digital educational system, the cost-effectiveness and security for online exam and remote teaching-learning can be ensured by VPN. VPN services have been offered in various forms over the years. Such services include ecommerce, social networking, and mobile networks.

Ecommerce: As VPN provides low-cost, secure connections, it becomes a popular technology for electronic commerce (ecommerce), which plays an important role in modern business world. Business-to-business (B2B) and business-to-consumer (B2C) are the two common types of ecommerce realization. B2B type of ecommerce holds the bigger portion. Banks like Chase Manhattan Bank and First Chicago National Bank have replaced their leased line networks with VPNs. Ecommerce allows electronic payment as a financial exchange online.

Virtual private social networks (VPSNs): Social networking sites (SNSs) are Internet-based applications that allow users to share information with a global audience. The biggest SNS, Facebook, has more than 500 million users. The concept of VPSNs is analogous to the concept of virtual private networks (VPNs) applied to traditional computer networks. VPSN could also store the real information in an encrypted way and decrypt them when necessary.

Mobile VPN: A virtual private mobile network (or mobile VPN) uses the physical infrastructure comprising a mobile wireless network as a platform to offer mobility as a service. Central to this network is a controller that accepts MVPN-related requests, such as creation and manipulation, and interacts with the infrastructure to fulfill them.

A VPN app is recommended for your mobile device to protect all your mobile communications. Most VPN services offer apps on both Android and iOS, saving you the trouble of configuring your phone's VPN's settings yourself.

Optical VPNs are the next step in the evolution of VPNs. It leverages the cost efficiency of optical networks.

7.5   Benefits and Challenges

We now weigh the pros and cons of VPNs. While the benefits of VPNs are many, their disadvantages are just as numerous. If used properly, a VPN is a technology with great opportunity for businesses of all sizes. It is cheaper than leased lines. It protects you from prying eyes. It prevents users on the Wi-Fi hot spot or any network from intercepting your traffic in a man-in-the-middle attack. Journalists and activists use VPN services to circumvent government censorship. VPNs are often used by journalists or people living in countries with restrictive policies on the use of the Internet.

The challenges typically facing VPN include the following:

1. *Cost:* There are hidden costs for deploying VPNs. For example, implementing encryption algorithms on a VPN is computer intensive and expensive. The cost of the network consists of the cost of running it (e.g., paying the administrators), the cost of renting the VPN channels or links over which the VPN tunnels are established, and the cost of the

core routers that serve as endpoints of the VPN.

2. *Complexity:* A problem with VPN is deciding how many VPN channels are required. The greater the number of channels, the more complex a VPN becomes. The number of endpoints per VPN is growing, and the communication pattern between endpoints is becoming increasingly complex and hard to predict. As a result, users are demanding dependable, dynamic connectivity between endpoints. For a large number of VPN endpoints, a tree structure scales better than point-to-point paths between VPN endpoint pairs.

3. *Incompatibility:* The VPN solutions and products that are offered by various vendors can have incompatibility issues and create complexity especially if these devices are not designed to work together. Several VPN implementations have not achieved full interoperability with each other.

4. *Administration:* The administration of VPNs require dedicated, skilled network administrators who know how a VPN works and are responsible for ongoing maintenance and management (such as monitoring, reporting, trouble shooting, etc.) of the network. The main concern for most enterprises is to find such competent network administrators who are usually expensive.

5. *Quality of service (QoS):* This ensures that all applications can coexist and function at acceptable levels of performance. Many time-sensitive applications need guarantees to function correctly. The VPN requires a certain quality of service (QoS), which is minimum guaranteed bandwidth for the connections between VPN sites.

6. *Malicious attacks:* Due to the vital role that VPNs play in corporate and governmental communication systems nowadays, they have been targeted for attacks. The private link supplied by a VPN can open a virtual backdoor to attackers. There are simple means of intercepting data passing through a network. Wi-Fi spoofing and Firesheep are two easy ways of hacking.

7.6   VPN Standards

For all this interconnectivity to work, there must be cooperation among vendors. The two largest groups are the Internet Engineering Task Force (www.ietf.org) and the Institute of Electrical and Electronics Engineers (www.IEEE.org). Each group has its own way of doing business and publishes its recommendations and standards. The IPSec suite of protocols was developed by IETF as part of IPv6. It is the security standard for the Internet, intranets, and VPNs. IPSec enables end-to-end security so that information sent to or from a computer can be secure.

Standards for VPN are yet to mature, but VPN Requests for Comments (RFCs) include the following:

- RFC 2637, PPTP, July 1999
- RFC 2685, VPNs Identifier, September 1999
- RFC 2735, NHRP Support for VPNs, December 1999
- RFC 2764, A Framework for IP Based VPNs, February 2000
- RFC 2917, A Core MPLS IP VPN Architecture. September 2000
- RFC 3809, Generic Requirements for PPVPN. June 2004
- RFC 3931, L2TPv3, March 2005
- RFC 4026, PPVPN Terminology, March 2005
- RFC 4031, Service Requirements for Layer 3 PPVPNs, April 2005
- RFC 4093, Problem Statement: Mobile IPv4 Traversal of VPN Gateways, August 2005
- RFC 4110, A Framework for Layer 3 PPVPNs, July 2005
- RFC 4111, Security Framework for PPVPNs, July 2005
- RFC 4176, Framework for L3 VPN Operations and Management, October 2005
- RFC 4265, Definition of Textual Conventions for VPN Management, November 2005
- RFC 4364, BGP/MPLS IP VPNs, February 2006
- RFC 4365, Applicability Statement for BGP/MPLS IP VPNs, February 2006
- RFC 4381, Analysis of the Security of BGP/MPLS IP VPNs, February 2006

- RFC 5251, Layer 1 VPN Basic Mode, July 2008
- RFC 5252, OSPF-Based Layer 1 VPN Auto-Discovery, July 2008
- RFC 5253, Applicability Statement for Layer 1 VPN (L1VPN) Basic Mode, July 2008
- RFC 5265, Mobile IPv4 Traversal across IPsec-Based VPN Gateways, June 2008

Summary

1. The VPN secures the private network using encryption to ensure that only authorized users can access the network and that the data cannot be intercepted.
2. The two common types of VPNs are site-to-site and remote access.
3. A VPN uses a public network (the Internet) for private communications and applies encryption to ensure privacy.
4. VPNs are characterized by their deployment (as remote access or site-to-site), tunneling, security, safety, topology, encryption, and protocols.
5. Types of VPNs include access VPN, intranet/extranet VPN, cloud VPN, unicast VPN, and multicast VPN.
6. Applications of VPNs include ecommerce, social networks, mobile VPN, and optical VPN.
7. Benefits of VPN include being cheaper than leased lines, protects from prying eyes, and prevents being intercepted.
8. Challenges include costs, complexity as the number of channels increases, incompatibility of products due to no standards, expensive administrative cost, and possibility of malicious attacks.

Review Questions

7.1 Should a VPN be regarded as a:
(a) LAN   (b) MAN   (c) WAN
7.2 VPN is a private network built on top of another existing private network.
(a) True   (b) False

7.3 One of the disadvantages of VPN is hidden
cost.

(a) True   (b) False

7.4 A VPN promises to deliver management
capabilities and performance properties comparable to a dedicated leased line network.

(a) True   (b) False

7.5 Which VPN tunneling protocol is being used
by non-IP-based networks?

(a) PP2P   (b) GRE   (d) L2TP   (c) IPSec

7.6 To implement a VPN may require the
following:

(a) Router   (b) Remote node   (c) Remote
access server   (d) All of these

7.7 This is not one of the challenges of VPNs:

(a) Flexibility   (b) Cost   (c) Administration
(d) Quality of service   (e) Incompatibility

7.8 VPN architecture needs to satisfy the following basic requirements except:

(a)   Performance   (b)   Security   (c)
Reliability   (d) Availability   (e) Scalability

7.9 Virtual private network (VPN) is:

(a) An intranet   (b) An extranet   (c) Both

7.10 VPNs do not yet have standards.

(a) Yes   (b) No

Answer: 7.1c, 7.2b, 7.3a, 7.4a, 7.5d, 7.6d, 7.7a,
7.8a, 7.9c, 7.10a

Problems

7.1 Why would an enterprise want to implement
a VPN?

7.2 Explain the terms of "virtual," "private," and
"network."

7.3 Explain the concepts of confidentiality, integrity, and authentication.

7.4 What is the difference between remote access
VPN and site-to-site VPN?

7.5 What is the major limitation of traditional
VPNs? How can this be overcome?

7.6 Why is encryption necessary in a VPN?

7.7 What are the pros and cons of using a VPN?

7.8 Describe two VPN tunneling protocols.

7.9 Discuss the security issues of VPN.

7.10 Discuss IPSec.

7.11 Write briefly about MPLS-VPN (RFC
2547).

7.12 What are the benefits of VPN?

7.13 Discuss three of the challenges of VPNs.

7.14 What components make up the cost of operating a VPN?

7.15 Describe two applications of VPN.

7.16 Write briefly about RFC 2685 and RFC
3809.

7.17 What is RFC 5252?

7.18 What is the significance of RFC 4381?

Digital Subscriber Line

8

Technology is nothing. What's important is that you have a faith in people, that they're basically good and smart, and if you give them tools, they'll do wonderful things with them.

Steve Jobs

Abstract

Digital Subscriber Line (DSL) being a high-speed data technology is referred to as a broadband telecommunication system. The availability and speed of the DSL service depend on the distance from the home or business to the closest telephone company facility. This chapter covers the different types of DSLs and their advantages and disadvantages.

Keywords

Digital Subscriber Line (DSL) · Asymmetric DSL (ADSL) · High-bit Data Rate DSL (HDSL) · Rate Adaptive DSL (RADSL) · ISDN DSL (IDSL) · Very High Data Rate DSL (VDSL)

8.1 Introduction

Digital Subscriber Line (DSL) is a wireline technology that provides the ability to transmit digital data over copper local telephone line infrastructure installed at homes and businesses. It is part of the access technology network that is supposed to fulfil the very fast data rate demands in a very growing market. DSL being a high-speed data technology is also referred to as a broadband telecommunication system. These DSL-based broadband systems provide transmission speeds ranging from several hundred Kbps to millions of bits per second (Mbps). Figure 8.1 shows the basic xDSL application block diagram.

There are different types of DSL, often referred to "xDSL" where the "x" stands for the various derivatives of DSL. The availability and speed of the DSL service depends on the distance from the home or business to the closest telephone company facility. Table 8.1 shows some of the different types of DSL systems. Each has different transmission speeds. The objective of this chapter is to discuss the different types of DSL such as Asymmetric DSL (ADSL), High-bit Data Rate DSL (HDSL), Rate Adaptive DSL (RADSL), ISDN DSL (IDSL), and Very High Data Rate DSL (VDSL). There are over 25 million xDSLs which is the second most popular platform in the United States that are currently in use, amounting to over 30 percent of the US broadband market.

8.2 ADSL

Asymmetric DSL (ADSL) is one of the types of broadband access technologies that belong to xDSL family. It supports voice, video, and data with over 60% market share being the best choice worldwide. It is called *asymmetric* because the

© The Author(s), under exclusive license to Springer Nature Switzerland AG 2022
M. N. O. Sadiku, C. M. Akujuobi, *Fundamentals of Computer Networks*,
https://doi.org/10.1007/978-3-031-09417-0_8

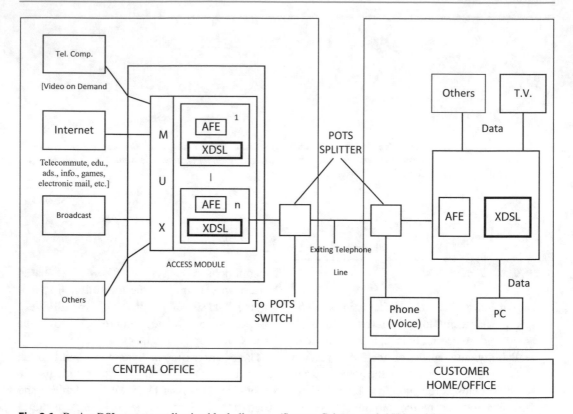

Fig. 8.1 Basic xDSL system application block diagram. (Source: Cajetan, et al. 2008, p. 166)

downstream data rate is from 1.5 to 8 Mbps while the upstream data rate is from 16 to 768 Kbps. This means that its data flows faster in one direction than the other. The data transmission has faster downstream to the subscriber than upstream. It is *digital* because no type of communication is transferred in an analog method. All data is purely digital and only at the end when it is modulated to be carried over the line. It is called a *subscriber line* because the data is carried over a single twisted-pair copper loop to the subscriber premises.

The ADSL and ADSL Lite Models and connectivity to TELCO CO and the Internet are shown in Fig. 8.2, while the ADSL System Model is shown in Fig. 8.3. The various components of these models include the ADSL modem at the customer premises (ATU-R), the modem of the central office (ATU-C), DSL Access Multiplexer (DSLAM), and the splitter which is an electronic low pass filter that separates the analog voice or

ISDN signal from ADSL data frequencies DSLAM. The ADSL requirements include phone line activated by whichever phone company you subscribe to for your ADSL service, filter to separate the phone signal from the internet signal, ADSL modem, and subscription with an ISP supporting ADSL.

8.2.1   ADSL Operation

ADSL exploits the unused analog bandwidth available in the wires. It works by using a frequency splitter device to split a traditional voice telephone line into two frequencies as shown in Fig. 8.4. The regular phone uses approximately about 4 kHz. The remaining frequency is used for the ADSL upstream from 25.875 KHz to 138 KHz and the downstream from 138 KHz to 110 KHz frequencies. Prior to the existence of DSL technology, these fre-

Table 8.1 Different types of DSL systems

DSL Type	Meaning	Data rate downstream speed	Data rate upstream speed	Distance	Applications
ADSL	Asymmetric DSL	1.5–8 Mbps	16–768 Kbps	Up to 4 km	Interactive multimedia, Internet access, remote office LAN residential and SOHO applications, video-on-demand
HDSL	High Data Rate DSL	T1: 1.544 Mbps E1: 2.048 Mbps	T1: 1.544 Mbps E1: 2.048 Mbps	Up to 5 km; Up to 12 km with repeaters	Telco transport applications, cellular base stations connectivity, T1/E1 leased lines
RADSL	Rate Adaptive DSL	1.5–8 Mbps	16 to 768 Kbps	Up to 6 km	Interactive multimedia, Internet access, remote office LAN residential and SOHO applications, video-on-demand
ISDL	ISDN DSL	144 Kbps	144 Kbps	Up to 4 km	Internet, intranet applications, etc.
VDSL	Very High Data Rate DSL	13–52 Mbps	1.5–6.0 Mbps	Up to 1.5 km	Full Service Access Network

Source: Cajetan, et al. 2008, p. 165

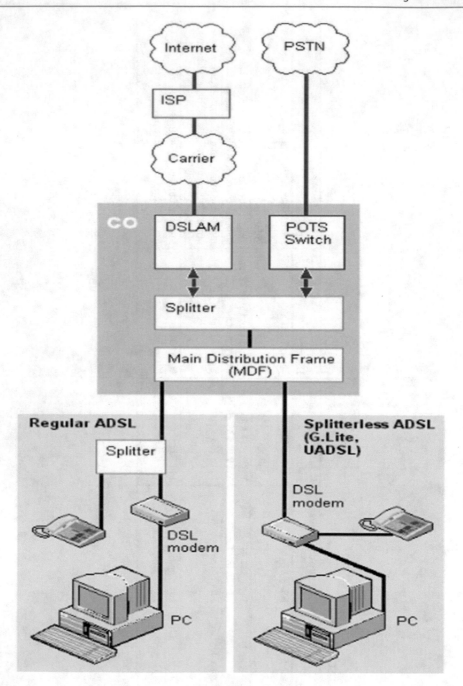

Fig. 8.2 ADSL and ADSL Lite models and connectivity to TELCO CO and the Internet. (Source: Cajetan, et al. 2008, p. 168)

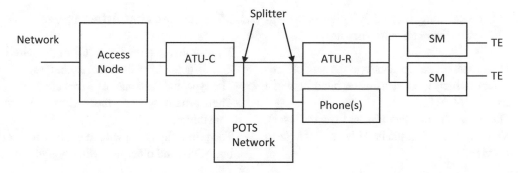

Fig. 8.3 ADSL System Model. (Source: Cajetan, et al. 2008, p. 168)

Fig. 8.4 How ADSL operates

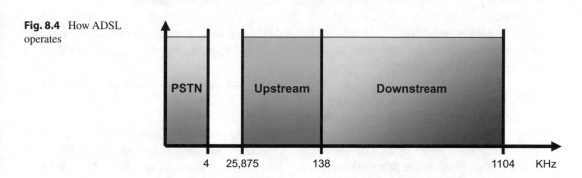

quencies were unused with the exception of the phone frequency of about 4 KHz.

There are two competing and incompatible standards for modulating the ADSL signal. They are:

- Carrierless amplitude phase (CAP)
- Discrete multitone (DMT)

8.2.2 ADSL Modulation Techniques

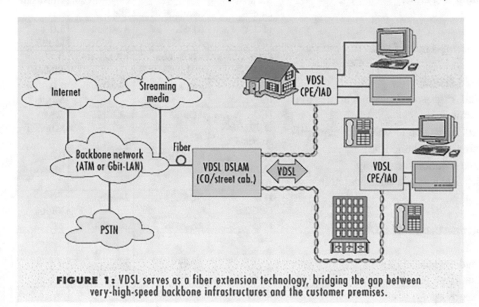

FIGURE 1: VDSL serves as a fiber extension technology, bridging the gap between very-high-speed backbone infrastructures and the customer premises.

CAP is an encoding method that divides the signals into two distinct bands. They are:

- The upstream data channel (to the service provider), which is carried in the band between 25 and 160 kHz
- The downstream data channel (to the user), which is carried in the band from 200 kHz to 1.1MHz

These channels are widely separated in order to minimize the possibility of interference between the channels.

DMT separates the DSL signal so that the usable frequency range is separated into 256 channels of 4.3125 kHz each.

- DMT has 224 downstream frequency bins (or carriers) and 32 upstream frequency bins.
- DMT constantly shifts signals between different channels to ensure that the best channels are used for transmission and reception.

This DMT modulation technique has become the standard for ADSL which combines quadrature amplitude modulation (QAM) and frequency division modulation (FDM).

8.2.3 Advantages of ADSL

- Fast download speeds
- Uninterrupted, high-speed Internet access that is always online
- Simultaneous Internet and voice/fax capabilities over a single telephone line
- Cost-effective solution for society
- Data security that exceeds other technologies

8.2.4 Disadvantages of ADSL

- Slower upload speeds
- Distance-sensitive
- Phone line required

8.2.5 Factors That Affect Speed

- The number and type of joints in the wire
- The distance from the local exchange
- The type and thickness of wires used
- The proximity of the wires to radio transmitters
- The proximity of the wire to other wires carrying ADSL and other non-voice signals

8.2.6 ADSL Standards

ADSL has several standards that govern its functions and operations as shown in Table 8.2.

8.3 HDSL

The High-bit-rate Digital Subscriber Line (HDSL) which was standardized in 1994 is an alternative to the T-1 line with 1.544 Mbps for both its downstream and upstream data rates. The T-1 line uses the Alternate Mark Inversion (AMI) encoding which is very susceptible to

Table 8.2 ADSL standards

Standard name	Common name	Downstream rate	Upstream rate
ITU G.992.1	ADSL (G.DMT)	8 Mbit/s	1.0 Mbit/s
ITU G.992.2	ADSL Lite (G.Lite)	1.5 Mbit/s	0.5 Mbit/s
ITU G.992.3/4	ADSL2	12 Mbit/s	1.0 Mbit/s
ITU G.992.3/4 Annex J	ADSL2	12 Mbit/s	3.5 Mbit/s
ITU G.992.3/4 Annex L	RE-ADSL2	5 Mbit/s	0.8 Mbit/s
ITU G.992.5	ADSL2+	24 Mbit/s	1.0 Mbit/s
ITU G.992.5 Annex L	RE-ADSL2+	24 Mbit/s	1.0 Mbit/s
ITU G.992.5 Annex M	ADSL2+	28 Mbit/s	3.5 Mbit/s

attenuation at high frequencies. This limits the length of the T-1 to 1 km which is about 3200 ft. A repeater is needed to boost the signals for longer distance which increases the cost of using T-1 lines.

HDSL uses 2B1Q encoding technology which is less susceptible to attenuation. It has a data rate of 1.544 Mbps and up to 2 Mbps without a repeater up to a distance of 3.86 km which is about 12000 ft. The HDSL was the first derivative of an xDSL to use a higher frequency spectrum of copper twisted-pair cables.

8.3.1 HDSL Operation

HDSL operates symmetrically because both its downstream and upstream data rates are the same about 1.544 Mbps. It can also operate in different formats such as:

(i) In full-duplex 2.048-Mbps transmission European (E1 speed) over three unconditioncd twisted-pair lines.

(ii) In full-duplex 1.544-Mbps transmission American (T1 speed) over unconditioned, unshielded twisted-pair cables. In this case, the twisted-pair phone lines have two pairs of wires (four wires) within the line. In full-duplex, it operates with 784-Kbps transmission taking place over each pair of wires. This is normally called the dual-duplex HDSL, and it is the most common configuration.

(iii) In full-duplex 668-Kbps transmission over a single unconditioned, unshielded copper twisted-pair phone line.

HDSL can be used for connections between points of presence (POPs), private data networks, and other services. It can operate for a carrier service area (CSA) from the telecommunication company's central office (CO) for a maximum distance of about 3700 meters. Different types of devices can be connected to the HDSL in its operation like routers, bridges, and telephone equipment such as private branch exchanges

(PBXs) over a campus using HDSL line drivers with built-in CSU/DSU (channel service unit/data service unit) functionality as shown in Fig. 8.5.

HDSL can be used to connect to campus networks and phone equipment at T1 speeds without the need for costly fiber-optic cabling. The line drivers or line terminals for HDSL operations generally support a variety of data interfaces, including 10BaseT, G.703, and V.35 connections. They are configurable for N x 64 Kbps transmission speeds and sometimes include bridge or router functionality for framing of Point-to-Point Protocol (PPP), Internet Protocol (IP), High-level Data Link Control (HDLC), and other protocols. They can be used for LAN-LAN connections and for connecting local area networks (LANs) to frame relay networks or the Internet as shown in Figs. 8.1 and 8.5. Because HDSL requires a repeater only at both ends of the line, not at every 1800 meters (6000 feet), as required by conventional T1 line, it makes HDSL easier to maintain and provision than conventional T-carrier span designs.

8.3.2 Different Varieties of HDSL

There are different types of HDSL such as e.SHDSL and G.SHDSL. e.SHDSL is the Ethernet version of HDSL which is over Copper, offering symmetrical bandwidth at more than 100 Mbps over multiple bonded copper pairs without repeaters. The G.SHDSL which can also be called G.991.2 is an international standard for symmetric DSL developed by the ITU. Its main function is to send and receive high-speed symmetrical data streams over a single pair of copper wires at rates between 192 kbps and 2.31 Mbps. The G.SHDSL was developed to incorporate the features of other DSL technologies, such as ADSL and SDSL, and transport T1, E1, ISDN, ATM, and IP signals. This is the first DSL technology to be developed from the ground up as an international standard. The standard for G.SHDSL was ratified by the International Telecommunication Union (ITU) in February 2001.

Fig. 8.5 HDSL
operational diagram

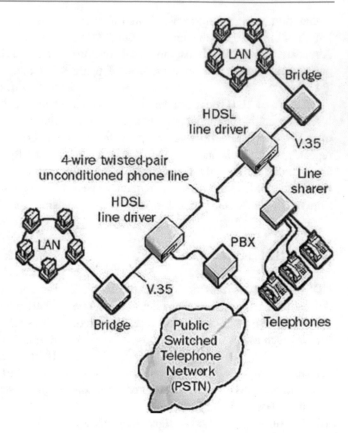

8.3.3   Difference Between HDSL
and ADSL

HDSL and ADSL are the different variants of the xDSL. High-bit-rate Digital Subscriber Line (HDSL) permits transmission up to 12 Kft from the central Digital Subscriber Line Access Multiplexer (DSLAM) or from the Central Office (CO). It supports telecommunication transport application and cellular base station connectivity. It has a downstream and upstream rate of about 1.544 Mbps. In the case of the ADSL, the downstream is from 1.5 to 8 Mbps, while the upstream is from 16 to 768 Kbps. The ADSL supports interactive multimedia, Internet access, remote office LAN residential and small office/home office (SOHO) applications, and video-on-demand. The maximum bandwidth for HDSL transmissions is less than the maximum bandwidth for ADSL. HDSL is not widely implemented at the customer premises level, which uses the more popular ADSL or G.Lite for providing customers with high-speed Internet access.

8.4   RADSL

The *Rate Adaptive Digital Subscriber Line* (RADSL) is one of the derivatives of the xDSLs technology that can enable telephone companies to simultaneously transmit up to 400 channels of digital television programming, traditional voice services, and high-speed Internet access using existing copper infrastructure as shown in Fig. 8.6 from branch sites 1, 2, and 3. RADSL supports downstream data rates of up to approximately 8 Mbit/s and upstream data rates of up to approximately 1 Mbit/s. It can coexist with the plain old telephone service (POTS) voice on the same line.

Because RADSL seeks for the fastest rate during its operation, it allows for rate adaptation while the connection is in operation. It should be noted also that rate adaptation while the connection is in operation is one of the options in ADSL2, ADSL2+, and VDSL2, under the name seamless rate adaptation (SRA).

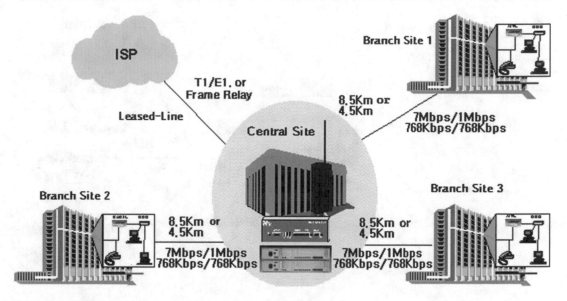

Fig. 8.6   RADSL operational diagram. (Source: Wikipedia, Google)

8.4.1   Difference Between RADSL and ADSL

The Rate Adaptive Digital Subscriber Line (RADSL) is a derivative of ADSL. However, it distinguishes itself from ADSL by having the capability to first test the line at startup of its operations and adapt its operating speed to the fastest the line can handle hence the name rate adaptive DSL. In the case of ADSL, first it is asymmetric which means the downstream and upstream data rates are not symmetric. It therefore plays a significant role for home use where the customers are more of consumers of data than producers of data. Therefore, a faster downstream (download) speed can be traded off for a slower upstream (upload) speed. Secondly, the ADSL standard allows for regular phone service to be implemented at the low end of the frequency spectrum. This means that the ADSL has a need for a splitter to be placed at the time of installation.

8.5   IDSL

IDSL is a derivative of the xDSL that is known as ISDN DSL, where ISDN means Integrated Service Digital Network (ISDN) Digital Subscriber Line. IDSL is defined as the system that transmits digital data at 128 Kbps on a regular copper telephone line (twisted pair) from a user to a destination using digital instead of analog or voice transmission as shown in Fig. 8.7. In this process of transmission, it bypasses the telephone company's central office equipment that handles analog signals. The only one possible technology in the Digital Subscriber Line approach that allows use of existing ISDN card technology for data-only use is the IDSL. In IDSL, the downstream and upstream data rates are 144 Kbps each. It can extend its reach up to 4 Km. It uses one pair of wires. Internet and intranet are some of the application areas of IDSL.

8.6   VDSL

The Very High-Bit-Rate Digital Subscriber Line (VDSL) technology provides up to 52 Mbps of digital bandwidth over a single pair of copper wire. This allows several channels of high-quality video streams to be simultaneously transmitted to a user's location without rewiring as shown in Fig. 8.8. It is a data communications technology that enables faster data transmission over copper telephone lines. VDSL-based products have been developed for the evolving applications that will

Fig. 8.7 IDSL
operational diagram.
(Source: https://www.
networxsecurity.org/
members-area/
glossary/i/isdn.html)

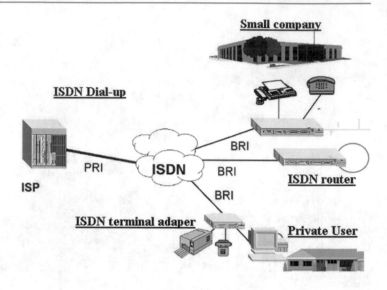

take our civilization into the next millennium. These applications allow us to increase the speed at which we communicate and learn. The key applications for the VDSL systems are:

- Internet/intranet applications
- Video-on-demand
- Telemedicine
- Videoconferencing
- Distance learning
- Electronic publishing
- Electronic commerce

Many of these applications will require more bandwidth than IDSL and even ADSL can supply. VDSL has been developed to enhance our ability to communicate using multimedia at high bandwidths.

8.6.1 Differences Between VDSL and ADSL

The key differences between VDSL and ADSL are as follows:

- The major difference between VDSL and ADSL technologies is speed. VDSL can reach speeds up to 52 Mbps for the downstream speed data rate and 16 Mbps for the upstream speed data rate. With ADSL, you can achieve a maximum downstream speed data rate of 8 Mbps and 1 Mbps as the upstream speed data rate.
- Due to the extremely high speeds supported by VDSL, it is seen as a good future technology for the use of applications that need a high bandwidth compared to ADSL.
- VDSL is faster than ADSL.
- While VDSL supports high-definition television (HDTV), ADSL does not support HDTV.
- VDSL allows adjustable bandwidths while ADSL cannot.
- VDSL has more trouble with damping compared to ADSL.
- VDSL is not very suitable for homes that are much further away from the telephone exchange compared to ADSL.
- VDSL is not widely distributed as compared to ADSL which is more widely distributed.

Summary

1. Digital Subscriber Line (DSL) is part of the access technology network that is supposed to fulfill the very fast data rate demands in a very growing market.
2. DSL being a high-speed data technology is also referred to as a broadband telecommunication system.

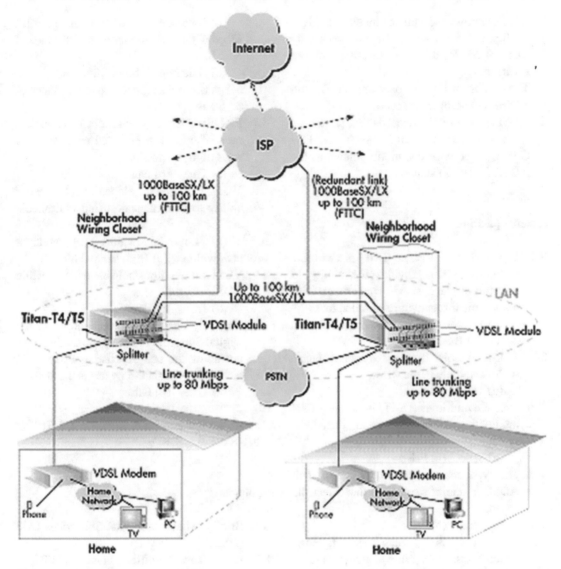

Fig. 8.8 VDSL operational diagram. (Source: https://www.americantechsupply.com/vdsl.htm)

3. ADSL is called *asymmetric* because the downstream data rate is from 1.5 to 8 Mbps while the upstream data rate is from 16 to 768 Kbps which means that its data flows faster in one direction than the other.

4. The high-bit-rate Digital Subscriber Line (HDSL) is an alternative to the T-1 line with 1.544 Mbps for both its downstream and upstream data rates.

5. HDSL operates symmetrically because both its downstream and upstream data rates are the same about 1.544 Mbps.

6. The G.SHDSL is the first DSL technology to be developed from the ground up as an international standard.

7. The maximum bandwidth for HDSL transmissions is less than the maximum bandwidth for ADSL.

8. Because RADSL seeks for the fastest rate during its operation, it allows for rate adaptation while the connection is in operation.

9. The only one possible technology in the Digital Subscriber Line approach that allows use of existing ISDN card technology for data-only use is the IDSL.

10. VDSL allows several channels of high-quality video streams to be simultaneously transmitted to a user's location without rewiring.
11. The major difference between VDSL and ADSL technologies is speed.
12. ADSL has several standards that govern its functions and operations such as ITU G.992.1 otherwise commonly known as the ADSL (G.DMT) standard.

Review Questions

8.1. Digital Subscriber Line (DSL) is a wireline technology that provides the ability to transmit digital data over copper local telephone line infrastructure installed to:
    (a) Homes and businesses   (b) Homes only   (c) Businesses only
8.2. ADSL exploits the unused analog bandwidth available in the wires.
    (a) True   (b) False
8.3. Discrete multitone (DMT) is not one of the modulation techniques used in ADSL systems.
    (a) True   (b) False
8.4. The type and thickness of wires used in ADSL is one of the factors that affect the speed.
    (a) True   (b) False
8.5. HDSL uses 2B1Q encoding technology which is less susceptible to attenuation.
    (a) True   (b) False
8.6. HDSL has:
    (a) Data rate of 1.00 Mbps and up to 1.2 Mbps without a repeater up to a distance of 3.86 Km

(b) Data rate of 1.00 Mbps and up to 1.2 Mbps without a repeater up to a distance of 3.86 Km
(c) Data rate of 1.544 Mbps and up to 2 Mbps without a repeater up to a distance of 3.86 Km
8.7. RADSL does not support downstream data rates of up to 8 Mbit/s and upstream data rates of up to 1Mbit/s.
    (a) True   (b) False
8.8. VDSL has been developed to enhance our ability to communicate using multimedia at:
    (a) Low bandwidths   (b) Medium bandwidths   (c) High bandwidths
8.9. VDSL allows adjustable bandwidths while ADSL cannot.
    (a) True   (b) False
8.10. The only one possible technology in the Digital Subscriber Line approach that does not allow the use of existing ISDN card technology for data-only use is the IDSL.
    (a) True   (b) False

Answer: 8.1 a, 8.2 a, 8.3 b, 8.4 a, 8.5 a, 8.6 c, 8.7 b, 8.8 c, 8.9 a, 8.10 b

Problems

8.1. Briefly, what is a Digital Subscriber Line (DSL)?
8.2. List at least five different types of DSL.
8.3. Complete the missing information in Table 8.3:
8.4. (a) Define Modulation. (b) What are the two types of modulation techniques that are applicable to the ADSL system?

Table 8.3 Different types of DSL systems

DSL type	Meaning	Data rate downstream speed	Data rate upstream speed	Distance	Applications
ADSL	Asymmetric DSL				
HDSL	High Data Rate DSL				
RADSL	Rate Adaptive DSL				
ISDL	ISDN DSL				
VDSL	Very High Data Rate DSL				

8.5. Briefly describe the CAP modulation system.

8.6. Briefly describe the DMT modulation system.

8.7. What are the advantages of ADSL?

8.8. What are the disadvantages of ADSL?

8.9. What are the factors that affect the speed of ADSL systems?

8.10. What is a HDSL? Describe briefly.

8.11. Describe briefly the different types of HDSL.

8.12. What are the differences between HDSL and ADSL?

8.13. Describe briefly the RADSL.

8.14. What are the differences between RADSL and ADSL? Describe briefly.

8.15. What is IDSL? Describe briefly.

8.16. What are the different applications of a VDSL?

8.17. What are the key differences between VDSL and ADSL?

Optical Networks

9

Success isn't about how much money you make; it's about the difference you make in people's lives.

Michelle Obama

Abstract

The chapter focuses on optical networks. Optics is the branch of physics that deals with light. An optical network is an efficient and promising infrastructure for handling the explosively growing IP traffic, increasing users' demand in terms of bandwidth. It connects devices using optical fibers and uses light waves to carry information in various types of communications networks. These impressive advantages of fiber optics over electrical media have made optical fiber to replace copper (twisted pair and coaxial cable systems) for fast transmission of enormous amounts of information. The properties make fiber optics attractive to government bodies, banks, and others with security concerns.

Keywords

Optics · Optical networks · Optical fibers · Fast transmission · High bandwidth

9.1 Introduction

Today, the Internet is expected to meet the growing demand of fast information exchange and bandwidth-intensive networking services. The increasing demand for bandwidth is driven primarily by IP traffic from video services, telemedicine, social networking, and mobile phones. The volume of data traffic transported across communication networks has grown rapidly and exceeded the volume of voice traffic. Communication has become multimedia involving voice, video, data, and images. This has driven the development of high-capacity optical networks. Optical network is an efficient and promising infrastructure for handling the explosively growing IP traffic, increasing users' demand in terms of bandwidth, and to their requirements in terms of quality of service (QoS). Since its invention in the early 1970s, the demand for and deployment of optical fiber have grown drastically.

Optics is the branch of physics that deals with light. It is gaining attention in various areas due to its usefulness in military, medicine, industry, space explorations, and telecommunications. Optical fiber has become the dominant transport medium in communication systems due to its inherent advantages in capacity, reliability, cost, and scalability. Optical networking will be the key in making the new world of communications possible. Optical networks (or photonic networks) are high-capacity communications network based on optical technologies and components. These networks are being equipped with ultra-long-reach technology and fiber-optic lines that can carry information thousands of kilometers before any signal regeneration is required.

Optical networks are classified according to the size of the area they cover:

- A *local area network* (LAN) links two or more points within a small area such within a building or campus, e.g., free-space optics.
- A *metropolitan area network* (MAN) covers a larger area such as a city, e.g., metro optical networks.
- A *wide area network* (WAN) extends over a larger area such as a nation or continent, e.g., optical Internet and SONET/SDH.
- An *access network* is a network that provides high-speed Internet connectivity to homes. It is the "last mile" of a telecom network connecting the telecom Central Office with end users. Access networks have different features and requirements than LANs, MANs, and WANs, e.g., Digital Subscriber Line (xDSL) and fiber to the x (FTTx).

An alternative classification used by the telephone industry refers to LANs as access networks, MANs as metro networks, and WANs as transport networks. Sometimes WANs are regarded as the core network (or the backbone infrastructure), while MANs and LANs are called edge networks.

An optical communication uses light waves to carry information in various types of communications networks. An optical network connects devices using optical fibers.

An **optical network** is a type of communication network that uses optical fiber cables as the primary communication medium.

Common optical networks are communication networks that include passive optical networks, free-space optical networks, and SONET/SDH. These networks are used in LAN, MAN, and WAN.

This chapter provides an introduction to optical networks. It begins with providing the reasons optical fiber are important for communication networks. Then it discusses some of the main optical components. It covers various optical networks: WDM-based networks, passive optical networks, SONET, all-optical networks, and free-space optics. It provides some applications of optical networks.

9.2   Why Optical Fiber?

Originally, communications network used copper to transfer information. In the mid-1970s, it was recognized that the existing copper technology would be unsuitable for future communication networks due to some inherent problems in copper wire: (1) its bandwidth is limited due to physical constraint; (2) it is susceptible to radio, electrical, and crosstalk interference which can garble data transmission; (3) its electromagnetic emissions compromise security; and (4) its reliability is reduced in the presence of a harsh environment. It was clear that light waves could have much higher bit rates than copper without crosstalk. Thus, fiber is regarded as the way of the future.

Optical networks are high-capacity communications networks based on optical technologies and components that provide routing, grooming, and restoration at the wavelength level as well as wavelength-based services. Several factors are driving the need for optical networks and making optical fiber the medium of choice. Fiber optical cables have the following advantages over copper cables.

- *High bandwidth*: Since the quantity of information which can be transmitted by electromagnetic waves increases in proportion to its frequency, by using light, four or five orders of magnitude can be gained in the amount of information being transmitted. Light has an information-carrying capacity (or bandwidth) 10,000 times greater than the highest radio frequencies. Therefore, optical fiber provides a very high capacity for carrying information because of the tremendous bandwidth available at optical frequencies (of the order of 40 THz). Because of this, optical networks can be regarded as telecommunications network of high capacity.

- *High transmission rate*: Fiber-optic transmits at speeds much higher than copper wire. Although the current Gbps network is regarded as a milestone compared with the existing Mbps networks, it is only a small fraction of the possible rates with fiber-optic technology. Fiber is capable of transmitting 3 TV episodes in 1 s and will be able to transmit the equivalent of an entire 24-volume encyclopedia in 1 s.
- *Attenuation:* It provides low attenuation and is therefore capable of transmitting over a long distance (up to 80 km) without the need of repeaters. This allows one to extend networks to campuses of large size or a city and its suburbs.
- *Electromagnetic immunity:* Fiber is a dielectric material. Therefore, it neither radiates nor is affected by electromagnetic interference (EMI), lightning strikes, or surges. The benefits of such immunity include the elimination of ground loops, signal distortion, and crosstalk in hostile environments.
- *Security:* Telecommunication companies need secure, reliable systems to transfer information between buildings and around the world. It is more secure from malicious interception because the dielectric nature of optical fiber makes it not easy to tap a fiber-optic cable without interrupting communication. Accessing the fiber requires intervention that is easily detectable by security surveillance. Fiber-optic systems are easy to monitor. Fiber-optic cables are virtually unaffected by atmospheric conditions. As the basic fiber is made of glass, it will not corrode or be affected by most chemicals. It can be buried directly in most kinds of soil or exposed to corrosive atmosphere without significant concern.
- *Cost:* Glass fibers are made from silica sand, which is more readily available than copper. The cost of optical fibers has fallen consider-

ably over the last few years and will continue to fall. So is the cost of related components such as optical transmitters and receivers.

- *Restoration capability:* As network planners use more network elements to increase fiber capacity, a fiber cut can have massive implications. By performing restoration in the optical layer rather than the electrical layer, optical networks can perform protection switching faster and more economically.
- *Wavelength services:* One of the great revenue-generating aspects of optical networks is the ability to resell bandwidth rather than fiber. By maximizing capacity available on a fiber, service providers can improve revenue by selling wavelengths, regardless of the data rate required. To customers, this service provides the same bandwidth as a dedicated fiber.

These impressive advantages of fiber optics over electrical media have made optical fiber to replace copper (twisted pair and coaxial cable systems) for fast transmission of enormous amounts of information from one point to another. These properties make fiber optics attractive to government bodies, banks, and others with security concerns.

However, optical fiber has its drawbacks. First, electrical-to-optical conversion at the sending end and optical-to-electrical conversion at the receiving end is costly. Second, optical fiber suffers the same fate as any wired medium—cable must be buried along the right of way. Third, special installation and repair techniques are required. Fourth, optical cable can only be used on the ground; it cannot work with mobile communication. Fifth, optical fiber is fragile and vulnerable to damage; you cannot twist or bend the cable. It is subject to frequent damage due mostly to fiber cuts.

9.3 Major Optical Components

As shown in Fig. 9.1, the three main components of an optical network are optical transmitter, fiber-optic cable, and an optical receiver. Other

Fig. 9.1 Basic optical network

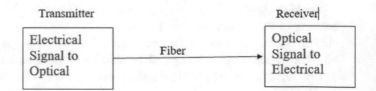

Transmitter

Electrical Signal to Optical

Fiber

Receiver

Optical Signal to Electrical

components include optical amplifiers, optical switches, optical routers, optical add/drop multipliers, optical splitters, multiplexer/demultiplexer, and optical crossconnects.

1. *Fiber-optic cables:* These are the most important component of optical networks. They carry light signals from node to node.

2. *Optical transmitter/receiver:* The optical transmitter converts an electrical signal received from a network node into light pulses, which are then placed on a fiber-optic cable for transmission. When a signal reaches the destination network, it is converted into an electrical signal by an optical receiver. Thus, a transmitter

An **optical fiber** is a very thin dielectric guide which carries signals in the form of light.

Optical fiber usually consists of a cylindrical core or filament of silica surrounded by cylindrical cladding, also of silica, with a lower refractive index, as shown in Fig. 9.2. A single fiber can transmit many separate signals simultaneously at different wavelengths of light. This technique is known as wavelength-division multiplexing (WDM). Since light beams do not interfere with each other, a single strand of fiber-optic cable can carry optical signals on multiple wavelengths simultaneously. There are three types of fiber-optic cables: single mode fiber, multimode fiber, and multimode graded index fiber.

converts data to a modulated optical signal, while the receiver converts a modulated optical signal to data. The most commonly used optical transmitter is LEDs (light-emitting diodes) and laser diodes. A common optical receiver is photodetector, which is the key part of an optical receiver.

3. *Optical amplifiers:* The optical fibers can transmit light waves more than 100 km without amplification, much farther than how long copper wires can transmit electronic signals. When the signal must span a longer distance, an optical amplifier is used to multiply the strength of the optical signal. Optical amplifiers are inserted in the line to compensate for fiber attenuation and therefore to increase the reach of transmission systems. The amplification is often achieved using erbium-doped fiber amplifiers (EDFA), which can easily amplify many different signals all at once.

4. *Optical switches:* Electronic switching is limited by its processing speed. Fiber-optic cables carry signals from node to node, with optical switches directing them to their destination. This switch opens and closes to direct the incoming and outgoing signals to the right place. When a signal arrives at its destination, it must be separated from the rest of the optical channels (or lightpaths). An optical network can be either switched or broadcast network. In switched networks, optical switches are used to route the optical signal. In broadcast optical networks, the

Example 9.1

The amount of fiber deployment by a telecom or city is often measured in *sheath* miles. Sheath miles is the total length of fiber cables, where each *route* in a network comprises many fiber *cables*.

For example, a 10-mile-long route using three fiber cables is said to have 10 route miles and 30 *sheath* (cable) miles. Each cable contains many *fibers*. If each cable has 10 fibers, the same route is said to have 300 fiber miles.

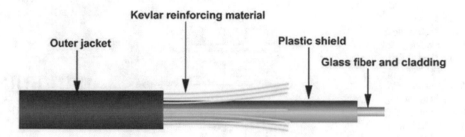

Fig. 9.2 A fiber-optic cable. (Source: "What is the fiber optic cable?" https://www.quora.com/What-is-the-fiber-optic-cable)

transmitted optical signals are received by all the nodes. Broadcast optical networks are much simpler to implement.

5. *Optical routers:* These direct each incoming optical signal to an appropriate outgoing fiber just like electronic routers. Optical routers in a packet-switched network make use of an optical label that is coded with the routing information such as the destination address. This label helps to transport packets across the core network in an all-optical fashion. This means that an IP packet is never converted into the electric domain until the packet arrives at an edge router.

6. *Optical add/drop multiplexers* (OADMs): These are devices that are used to add or remove single wavelengths from a fiber without disturbing the transmission of other signals. In other words, an OADM is a device that enables a number of wavelength channels to be dropped or added locally.

7. *Optical splitter:* An optical splitter is a passive component that is used in the optical network system to achieve the optical coupling, branching, and distributing. It is a device with multiple import or output terminals. Generally, we use M x N to indicate a splitter with M import and N output numbers. The splitter is the key technology that allows the access network to be electrically passive and is used in FTTH passive optical network.

8. *Multiplexer/demultiplexer:* These are also called mux/demux. Multiplexing is necessary because it is more economical to transmit data at higher rates over a single fiber than it is to transmit at lower rates over multiple fibers. As data rates get higher and higher, it becomes harder for electronics to process data. There are three common multiplexing techniques:

(a) Time-division multiplexing (TDM) separates signals by interleaving bits, one after another.

(b) Wavelength-division multiplexing (WDM) multiplexes a number of signals onto a single optical fiber by using different wavelengths (i.e., colors).

(c) Frequency-division multiplexing (FDM) separates signals by modulating the data onto different carrier frequencies.

The implementation of FDM in the optical domain requires multiple transmitters operating at different wavelengths. Therefore, optical FDM is known as wavelength-division multiplexing (WDM). Currently, WDM is preferred to TDM technique due to the complexity of the hardware required for TDM. Figure 9.3 shows TDM and WDM.

The multiplexer multiplexes the wavelength channels. The demultiplexers separate the optical channels and distribute them to separate optical receivers. A multiplexer combines multiple optical signals, while a demultiplexer separates signals having different carrier wavelengths. Typically, wavelength multiplexing is performed at the central office, while wavelength demultiplexing is provided at the customer's house.

9. *Wavelength-division multiplexer:* The technology of using multiple optical signals on the same fiber is called wavelength-division multiplexing (WDM). A WDM optical network divides the vast transmission bandwidth available on a fiber into several different smaller capacity "channels." In WDM signals, we are interested in the wavelength λ or frequency f of the signals, which are related by the equation

$$c = f \lambda$$

Fig. 9.3 Two
multiplexing techniques:
(**a**) TDM, (**b**) WDM.
(Source: "Optical
networks," in G. Pujolle
(ed.), *Management,
Control and Evolution of
IP Networks.* Newport
Beach, CA: ISTE Ltd.,
Chapter 26, 2007,
p. 599)

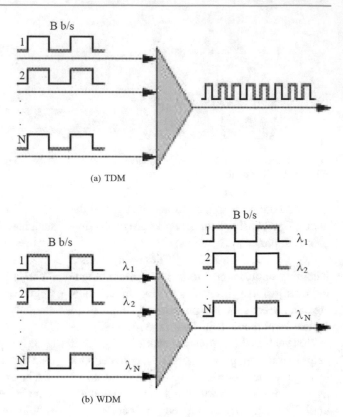

where $c = 3 \times 10^8$ m/s is the speed of light in free space (or vacuum). To characterize a WDM signal, we can use either its frequency or wavelength.

Example 9.2

The speed of light in fiber is actually somewhat lower than c. It is roughly 2×10^8 m/s. The wavelengths of interest in optical fiber communication are centered around 0.8, 1.3, and 1.55 μm, which lie in the infrared band that is not visible to the human eye. Using $c = 3 \times 10^8$ m/s, a wavelength of 0.8 μm would correspond to a frequency of

$$f = c/\lambda = \left(3 \times 10^8\right)/\left(0.8 \times 10^{-6}\right)$$
$$= 375 \times 10^{12} = 375\,\text{THz}$$

10. *Optical crossconnects* (OXCs): Their main function is to provide interconnection to a number of IP subnetworks. An optical crossconnect switches optical signals from input ports to output ports. It is a switching device capable of multiplexing/demultiplexing signals in both the wavelength and space domains. Each OXC can switch the optical signal coming in on a wavelength of an input fiber link to the same wavelength in an output fiber link. The OXCs provide the switching and routing functions for supporting the logical data connections between client subnetworks. There is interest in optical networks in which OXCs provide the switching functionality. An optical network often consists of interconnected optical subnetworks, where a subnetwork consists of OXCs built by the same vendor. An OXC can serve as an optical add/drop multiplexer (OADM), i.e., it can terminate the signal on a number of wavelengths and insert new signals into these wavelengths.

Other components of an optical network include optical coupler, circulator, wavelength converters, and transponder. Improvements in these components have led to an increase in the capacity of optical networks.

will take place. The dream of optical Internet and many other large-scale telecom networks supporting future bandwidth demands could only be made a reality with next-generation all-optical network.

9.4   WDM-Based Networks

With the advent of the optical amplifier and WDM technology (with hundreds of wavelengths per fiber), optics is playing a larger role in extending further to the edge of the network. The use of wavelength-division multiplexing (WDM) allows aggregation of many channels (or wavelengths) onto a single fiber. It is the simplicity of this multiplexing scheme which is the underlying reason that WDM technology is a key technology in the development of optical networks. Point-to-point WDM transmission is currently the dominant architecture for optical networks. This is the wavelength/optical analog of present electrical TDM-based networks. WDM is conceptually the same as frequency-division multiplexing (FDM), which is used in microwave radio and satellite systems. WDM transmission is regarded as a best candidate because it allows multiple optical signals at different wavelengths to transmit simultaneously along a single piece of optical fiber.

9.5   Passive Optical Networks

Most optical networks contain both active and passive optical elements. Active components can be located at the central office, customer's premises, in the repeaters, and in switches. Active elements require electrical power to perform their functions.

There are two types of optical networks: active optical networks and passive optical networks. Active optical networks use active components such as fiber amplifiers, while passive optical networks (PON) use only passive components such as optical filters and splitters. An active optical system uses electrically powered switching equipment, such as a router or a switch aggregator, to manage signal distribution and direct signals to specific customers. In such a system, a customer may have a dedicated fiber running to his or her house. The reliance of active optical networks on Ethernet technology makes interoperability among vendors easy. Because it requires

A **WDM optical network** consists of optical switches interconnected by fiber links, each of which allows aggregation of many channels onto a single fiber.

In a WDM optical network, each switch switches incoming optical packets to their desired output ports. WDM channels can be added easily to an existing WDM link. WDM is the technology which has cemented optical communication's inevitable widespread deployment. A

power, an active optical network inherently is less reliable than a passive optical network.

Passive optical networks (PONs) have become more popular and attractive due to their longevity, low operational costs, and huge bandwidth. For this reason, only PON will be considered here.

A **passive optical network** (PON) is a network that adopts a point-to-multipoint architecture and uses passive elements such as optical fiber and splitters, with no electrical powering required.

WDM network may offer data transparency in which the network may carry signals of arbitrary format. As time progresses IP over WDM

A PON is physically based on a tree topology, where a single central office is communicating with multiple geographically dispersed end users.

Fig. 9.4 A typical
passive optical network
(PON)

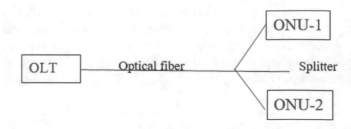

The optical fiber and splitters are the truly "passive" building blocks of the PON, with no electrical powering required. Powered equipment is required only at the source and receiving ends of the signal. As shown in Fig. 9.4, multiple users can be connected to a network element called the optical network unit (ONU), which in turn connects to a central office (CO) interface called an optical line terminal (OLT) through an optical fiber. The OLT is located at the CO, while ONUs can be located at the end user site.

The three popular PON alternatives are broadband PON (BPON), Ethernet PON (EPON), and gigabit PON (GPON). We will consider only EPON here. The widespread use of Ethernet makes it an attractive alternative transport technology for access networks. Instead of connecting one fiber from each user to the CO, the IEEE 802.3ah Task Force First Mile (FFM) developed the Ethernet passive optical network (EPON), where only optical cable is required between the CO and a set of 16 or 32 users. Thus, EPON is an extension of the IEEE 802.3 Ethernet LAN standard to fiber-optic access networks. EPON takes advantage of inexpensive and ubiquitous Ethernet equipment. This way fiber replaces copper (and DSL) in the access line, and enormous bandwidth is made available to users with connections to the backbone networks. Thus, PONs are widely deployed in the first/last mile of today's operational access networks. The simplified cabling infrastructure (with no active elements and less that can go wrong) has made PON an ideal and efficient for home Internet, college campuses, business environments, and voice and video applications. A typical PON deployment would involve a maximum reach of 20 km with a split of up to 64 and transmission speed of 1 Gbps.

Advantages of PONs include efficient use of power, simplified infrastructure and ease of upgrade, efficient use of infrastructure, and ease of maintenance. Disadvantages include distance,

test access, and high vulnerability to breakdown in the feeder line.

Example 9.3

WDM-PON is proving to be the most promising long-term, scalable solution for delivering high bandwidth to the end user. The key feature of PON operation is wave division multiplexing (WDM), used to separate data streams based on the wavelength (color) of the laser light. A PON based on WDM technology is shown in Fig. 9.5. The following different access networks can be obtained depending on the penetration of the optical fiber in these networks.

Solution

The term FTTx refers to "fiber to the x," where x could refer to one of several popular deployment scenarios.

FTTH (fiber to the home) is a passive optical network which is deployed all the way to the final subscriber.

FTTB (fiber to the business) is a passive optical network which is deployed all the way to the business/industrial center for providing information.

FTTCab (fiber to the cabinet) architecture uses the optical fiber until an electric cabinet located at approximately one kilometer from the final client.

FTTCurb (fiber to the curb) architecture increases the penetration of optical fiber in access networks. FTTCurb network makes it possible to serve a dozen clients.

FTTEx (fiber to the exchange) delivers enough necessary bandwidth for the new wideband services. The architecture serves roughly a thousand lines.

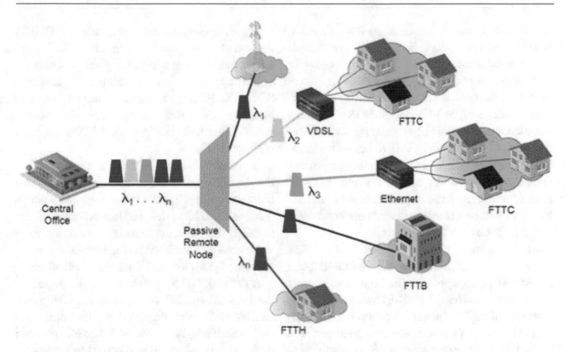

Fig. 9.5 Passive optical network based on WDM. (Source: "Fiber optic solutions," http://www.fiber-optic-solutions. com/overcome-challenges-wdm-pon-fttx.html)

9.6   SONET

There are two generations of optical networks. In the first generation, all the switching and other intelligent network functions were handled by electronics. Examples of first-generation optical

open standard optical interface for transmission at the broadband user-network and between network nodes. It is intended to provide a common international rate structure and eliminate the different transmission schemes and rates of nations.

SONET is an ANSI (American National Standards Institute) standard that defines a high-speed digital hierarchy for optical fiber.

networks are SONET (synchronous optical network) and the essentially similar SDH (synchronous digital hierarchy) network. The second generation of optical networks appeared in the 1980s. In these networks, routing, switching, and intelligence are carried out in the optical domain since electronics processing becomes more difficult at higher data rates. SONET in North America and SDH in Europe and Japan are the standard designs for fiber-optic networks. We consider just SONET in this section.

SONET is a fiber optics-based network for use by telephone companies. It provides an

An international version of SONET is the synchronous digital hierarchy (SDH). Voice traffic is carried on these fiber links using SONET (synchronous optical network)/SDH (synchronous digital hierarchy).

Three primary needs have driven the development of SONET. First, there was the critical need to move multiplexing standards beyond the DS3 (44.7 Mbps) level. Second, there was the need to enable cost-effective access to relatively small amounts of traffic with the gross payload of an optical transmission. Third, SONET has been motivated by data communication needs such as

LAN traffic, video, graphics, and multimedia. SONET provides the infrastructure over which broadband-integrated services can the deployed.

SONET is not a service but a transport interface that enables a communication network to carry various types of services. SONET is also not a communications network in the same sense as LAN, MAN, or WAN. It is rather an underlying distribution channel over which communications optical networks can all operate. SONET then provides the basis for the optical infrastructure that will be necessary to meet the communications of the decade and beyond.

A synchronous signal comprises a set of bytes organized into a frame structure. In SONET, the basic unit of transport is called the *synchronous transport signal level 1* (STS-1) frame, which has a bit rate of 51.84 Mbps and repeats every 125 μs, i.e., 8000 SONET frames are generated per second. An STS-1 frame structure is generally presented in the characteristic matrix shape in which every element represents a byte. Besides the STS-1, which serves as the basic building block, there are higher rate SONET signals (STS-N). An STS-N channel rate is obtained by synchronously mutiplexing N STS-1 inputs (i.e., byte interleaving N STS-1 signal together). Standards already define rates from STS-1 (51.84 Mbps) to STS-48 (2.48832 Gbps), as shown in Table 9.1, where

Synchronous optical networking (SONET) and synchronous digital hierarchy (SDH) have evolved as the most commonly used protocols for optical networks. One of the main features of a SONET/SDH rings is that they are self-healing, meaning that a ring can automatically recover when a fiber link fails. A typical SONET system is shown in Fig. 9.6.

In North America, SONET is fast becoming the de facto standard for high-speed broadband communications used by corporations, phone companies, universities, and others. Although the SONET/SDH framing structure is widely used in voice transmission in optical networks, it is cumbersome for high-speed links. The optical transport network (OTN) protocol was developed by the International Telecommunication Union as a successor. It was designed to transport data packet traffic such as IP and Ethernet over fiber optics in general and SONET/SDH in particular.

9.7   All-Optical Networking

Current networks combine optical networking and electronic networking. WDM plays a major role in optical networking and is currently the main mechanism for realizing all-optical networks (AONs), also called transparent networks.

An **all-optical network** (AON) has the user-network interfaces that are optical and data is carried from its source to its destination in optical form, without the need for optical-to-electrical conversion.

optical carrier level N (OC-N) is the optical equivalent of an STS-N electrical signal.

Table 9.1 SONET signal hierarchy

Electrical signal	Optical signal	Data rate (Mbps)
STS-1	OC-1	51.84
STS-3	OC-3	155.52
STS-9	OC-9	466.56
STS-12	OC-12	622.08
STS-18	OC-18	933.12
STS-24	OC-24	1244.16
STS-36	OC-36	1866.24
STS-48	OC-48	2488.32

AONs are attractive because they promise very high rates, flexible switching, and broad application support. All-optical WDM networks are fast becoming the natural choice for future backbone networks with huge capacities. The requirement of optical communication supporting large scalable networks in the future led to the concept of all-optical networks.

The original goal of the all-optical network was based on keeping the data signals entirely in the optical domain from source to destination. This eliminates the so-called electronic bottleneck, to allow arbitrary signal formats, bitrates, and protocols to be

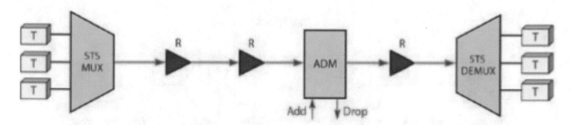

Fig. 9.6 A typical SONET system. (Source: "SONET: Synchronous optical network," https://pt.slideshare.net/hamza-sajjad9081/sonet-synchronous-optical-networking)

transported. An important aspect in AONs is transparency. This means that the data flows through the network without being interpreted in switches or routers. This also implies that any type of traffic could be simply passed through the network. Transparency brings advantages of security and ease in network upgrades. AONs could use WDM or TDM, but WDM optical network is more common. They can be applied to LAN, MAN, and WAN.

9.8   Free-Space Optics

Radio and fiber are dual technologies: radio provides mobility to the end user but has severe

WDM. This is not possible using any fixed wireless/RF technology existing today. Second, FSO technology requires no FCC licensing or municipal license approvals and thus obviates the need to buy expensive spectrum. This distinguishes it clearly from fixed wireless technologies. Third, FSO is essentially the same as that for fiber-optic transmission, and it uses lasers or LEDs. The only difference is the medium; the signal is sent through air or free space. Light travels through air faster than it does through glass or optical fiber. So one may regard FSO as optical communications at the speed of light. Figure 9.7 shows a typical free-space optical system.

Free-space optical (FSO) communications is a technology that can provide high-speed, cost-effective, and fast wireless networks that transmit at the speed of light.

bandwidth limitations; fiber has severely limited pervasiveness but brings enormous bandwidth potential. Free-space optics (FSO), also known as fiber-free or fiberless photonics, refers to the transmission of modulated light pulses through free space (air or the atmosphere). Laser beams are generally used, although non-lasing sources such as light-emitting diodes (LEDs) or IR-emitting diodes (IREDs) will serve the same purpose. FSO can be the best wireless solution where fiber optical cable is not available, high bandwidth (anywhere from 1 Mbps up to 1.25 Gbps) is required, and line-of-sight can be obtained to a target within a couple of miles.

FSO is an optical technology and not a wireless technology for three basic reasons. First, FSO enables optical transmission at speeds of up to 2.5 Gbps and in the future 10 Gbps using

FSO's greatest success so far has come from the LAN/campus connectivity market. It can also solve the "last mile" problem of connecting to fiber infrastructure in metropolitan areas. It operates in a completely unregulated frequency spectrum (range of THz), implying that FSO is not likely to interfere with other transmissions. FSO is fairly difficult to intercept because its beams are invisible, narrow, and very directional (aimed at a particular antenna). It offers a cost-effective, quick, and available infrastructure that is not only easily deployed (within days), redeployed, and easy to manage but can also offer a multitude of options—distance, speed, topology, and installations. It is now a viable choice for connecting LAN, WAN, and MAN and carrying voice, video, and data at the speed of light.

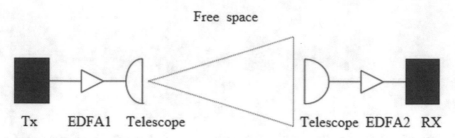

Free space

Fig. 9.7 A typical free-space optical system

9.9   Applications

With fiber optics, high-speed communication systems can be developed to offer services that are unprecedented such as in video gaming systems. The following are a few important applications that will be enabled by high-speed optical networks.

- *Internet and web browsing:* Bandwidth requirements for each user of the Internet have significantly increased, and the number of new users is rapidly growing.
- *Graphics and visualization:* The class of graphics and visualization applications is data- and compute-intensive. The distribution of images for medical consultation to multiple locations will require high-speed networks. This capability is critical in order to realize full telemedicine support.
- *Multimedia conferencing:* The rapid introduction of video conferencing has reduced business costs and travel time by bringing parties together over audio-video channels. However, realization of the full benefits of multimedia conferencing to support distant collaborative meetings depends on the availability of high-speed networks.
- *Broadband services to the home:* Telephone companies and cable TV operators are attempting to offer innovative new services to their residential and small business customers. The services may include video-on-demand, interactive TV, distance learning, and electronic commerce. Optical networking has been identified as the key technology to bring such broadband services to the home. For

example, optical access networks can be used for fiber to the home (FTTH).
- *Business:* For business organizations, the availability of high-capacity, low-cost networking promises to address today's business needs as well as effectively eliminate geographic boundaries for the business systems of the future. By providing unprecedented low-cost telecommunications capacity, all-optical networks have the potential to address these critical business issues.
- *Undersea networks*: The underlying principles of optical fiber communications for inland applications are somewhat applicable to undersea and terrestrial systems. However, stringent reliability requirements must be imposed on the undersea networks because of the need to minimize the number of repairs during the life of the cable. The cable must have sufficient mechanical strength to survive the harsh environment and the deep-sea cable-laying operations without damage.

Summary

1. Optical networking is currently the only technology that is flexible enough to handle the increasing dominance of data traffic.
2. An optical network is a communication network that connects devices using optical fiber cables.
3. An optical fiber cable is a cylindrical filament which makes it possible to conduct light over a long distance. Its high bandwidth and low attenuation characteristics make it ideal for high-speed, long-distance transmission.

4. Major components of optical networks include optical fiber cables, optical transmitter, optical receiver, optical amplifiers, optical switches, optical routers, optical add/drop multipliers, optical splitters, multiplexer/demultiplexer, and optical crossconnects.

5. A WDM (wavelength-division multiplexing) optical network consists of optical switches interconnected by fiber links, each of which allows aggregation of many channels onto a single fiber. WDM is the technology of combining multiple wavelengths onto the same optical fiber.

6. A passive optical network (PON) is a network that adopts a point-to-multipoint architecture and uses passive elements such as optical fiber and splitters, with no electrical powering required.

7. SONET (synchronous optical network) is an ANSI standard that defines a high-speed digital hierarchy for optical fiber. It is intended to provide a common international rate structure and eliminate the different transmission schemes and rates of nations.

8. Free-space optical (FSO) communications is a technology that can provide high-speed, cost-effective, and fast wireless networks that transmit at the speed of light.

9. An all-optical network (AON) has the user-network interfaces that are optical, and data is carried from its source to its destination in optical form, without the need for optical-to-electrical conversion.

10. Common applications of optical networks include web browsing, graphics, and visualization, multimedia conferencing, broadband services to the home, business, and undersea networks.

9.2. The light source used in fiber-optic communication is:
(a) LEDs and lasers (b) Xenon lights (c) Incandescent

9.3. Which of these components is not used in optical networks?
(a) Amplifier (b) Router (c) Multiplexer (d) RFID

9.4. Which of these optical networks is not a PON?
(a) Narrowband PON (NPON) (b) Broadband PON (BPON) (c) Ethernet PON (EPON) (d) Gigabit PON (GPON)

9.5. These components are passive:
(a) Amplifier (b) Switches (c) Splitters (d) Optical fiber

9.6. Which of these networks is used for providing information to the industrial center?
(a) FTTH (b) FTTB (c) FTTCab (d) FTTCurb

9.7. In SONET, the rate for STS-1 level of electrical signaling is:
(a) 2488.320 Mbps (b) 622.080 Mbps (c) 155.52 Mbps (d) 51.84 Mbps

9.8. Which of these is not an application of fiber optical networks?
(a) Web browsing (b) Visualization (c) Conferencing (d) Satellite services

9.9. SDH was developed by:
(a) ANSI (b) OSI (c) ITU (d) IEEE

9.10. Free-space optics (FSO) is a wireless technology.
(a) True (b) False

Answer: 9.1 (b), 9.2 (a), 9.3 (d), 9.4 (a), 9.5 (c, d), 9.6 (b), 9.7 (d), 9.8 (d), 9.9 (c), 9.10 (b)

Review Questions

9.1. Which of the following is not an advantage of using fiber optical cable?
(a) High-speed capability (b) High attenuation (c) Low signal distortion (d) Low power requirement (e) Low material usage

Problems

9.1. What led to the development of optical networks?

9.2. What are the problems with copper wire that optical fiber is meant to solve?

9.3. What is an access network?

9.4. Mention the advantages of optical networks.

9.5. What are the drawbacks of using optical fiber?

9.6. Explain the following terms in your own words: optical amplifiers, optical switches, and optical routers.

9.7. Describe multiplexer/demultiplexer.

9.8. Assume that an optical fiber transmits at the speed of 2×10^8 m/s. Calculate the wavelength at the operating frequency of 50 THz.

9.9. Describe WDM. Why does it play a major role in the development of optical networks?

9.10. What are the functions of optical crossconnects?

9.11. Describe EPON.

9.12. Do some research on BPON and GPON.

9.13. What are the common applications of PON?

9.14. What are the major advantages and disadvantages of PON?

9.15. Explain the difference between FTTH and FTTB.

9.16. Describe the needs that led to the development of SONET

9.17. Describe SONET.

9.18. Do some research on SDH.

9.19. What is an all-optical network (AON)? Why is it important?

9.20. Describe what transparency means in an AON.

9.21. What is free-space optics FSO?

9.22. Why is FSO regarded as an optical technology and not a wireless technology?

9.23. Mention some of the applications of high-speed optical networks.

Wireless Networks

10

Don't just learn, experience.
Don't just read, absorb.
Don't just change, transform.
Don't just relate, advocate.
Don't just promise, prove.
Don't just criticize, encourage.
Don't just think, ponder.
Don't just take, give.
Don't just see, feel.
Don't just dream, do.
Don't just hear, listen.
Don't just talk, act.
Don't just tell, show.
Don't just exist, live.

Roy T. Bennett

Abstract

In this chapter, we cover wireless networks, which allow a more flexible communication model than wired networks. Wireless communication uses electromagnetic or radio waves for transmission and reception. The wireless communication revolution is bringing fundamental changes to data networking. Wireless networking plays an important role in civil and military applications. It is invading our lives through the increasing use of smartphones and portable computers. It is common today to see people everywhere and at anytime use mobile telecommunications devices, such as cellular/mobile phones, personal digital assistants (PDAs), or laptops. Wireless technologies in use today include Wi-Fi, Bluetooth, ZigBee, cellular networks, wireless sensor networks, and more.

Keywords

Wireless networks · Wi-Fi · Bluetooth · ZigBee · Cellular networks · Wireless sensor networks

10.1 Introduction

As technology advances the need for wired and wireless networking has become essential in today's organizations. To support transmission of voice, data, and video, several wired and wireless communication network infrastructures have evolved throughout the past century. We have covered wired networks such as Ethernet and optical networks in previous chapters. In this chapter, we focus on wireless networks, which allow a more flexible communication model than wired networks.

M. N. O. Sadiku, C. M. Akujuobi, *Fundamentals of Computer Networks*,
https://doi.org/10.1007/978-3-031-09417-0_10

Wireless communication (or communication using radio waves) was invented by an Italian physicist, Guglielmo Marconi, in 1895. Since then, the worldwide growth of the wireless communications industry has been phenomenal. The evolution of the wireless communication has radically changed the nature and socialization of human communications. The wireless communication revolution is bringing fundamental changes to data networking. It is the fastest growing field in the telecommunication industry. Wireless networking plays an important role in civil and military applications. It is invading our lives through the increasing use of smartphones and portable computers. It is common today to see people everywhere and at anytime use mobile telecommunications devices, such as cellular/mobile phones, personal digital assistants (PDAs), or laptop.

Portable mobile devices, such as smartphones, that allow instantaneous high-speed information access have radically changed our

and also create mobility for devices connected to the network.

There are several types of wireless networks. The following seven are the most popular:

- *Wireless local area network* (WLAN): This links two or more devices over a short distance using a wireless means. Computers are often connected to networks using wireless link such as WLANs.
- *Wireless metropolitan area network* (WMAN): This wirelessly connects several wireless LANs.
- *Wireless wide area network* (WWAN): It covers large areas such as neighboring towns and cities.
- *Wireless personal area network* (WPAN): This interconnects devices in a short span, generally within a person's reach.
- *Cellular network:* This is a network distributed that employs a large number of low-power wireless transmitters to create the cells,

A **wireless network** is a computer network that uses wireless connections between network devices.

way of life. Wireless networks are networks that enable one or more devices to communicate without physical connections, without using wires or cables. They offer mobility and utilize radio waves or microwaves to maintain communication. Thus, the basis of wireless systems is electromagnetic or radio waves. (The physical laws that make radio possible are known as Maxwell's equations, identified by James Clerk Maxwell in 1864.) This wireless connectivity takes place at the physical layer of the network. When you use a wireless network, you trade wired connectivity for connectivity delivered via a radio signal.

Wireless networking avoids the costly process of using cables to connect devices in a building. Wireless networks offer many advantages when it comes to difficult-to-wire areas or trying to communicate across a street or river. Wireless technology is an effective option compared to Ethernet for sharing printers, scanners, and high-speed Internet connections. Wireless networks help save the cost and time of installation of cable

which are base stations transmitting over small geographic areas.
- *Satellite network*: This uses satellites to provide instant global connectivity and communications from anywhere to anywhere.
- *Wireless sensor network* (WSN): This usually consists of a large number (hundreds or thousands) of sensor nodes deployed over a geographical region.

Only these wireless networks will be covered in this chapter. While WLAN, WMAN, and WPAN are basically data networks, cellular network is for voice. The relationship between these wireless networks is shown in Fig. 10.1. Other wireless networks include wireless PBXs, wireless USB, WiMobile, HIPERLAN (in Europe), and ad hoc networks.

This chapter provides a brief overview of wireless communication networks. It begins with the basic fundamentals of wireless networks. Then it considers wireless local area networks (WLAN): the IEEE 802.11 standard. It covers

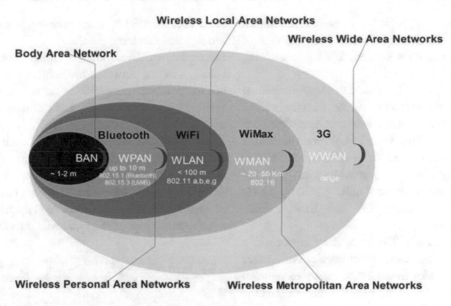

Fig. 10.1 Classification of wireless networks. (Source: Alomari, et al. 2011)

wireless metropolitan area networks (WMAN), wireless wide area network (WWAN), and wireless personal area network (WPAN). It explains cellular network and satellite networks. It considers wireless sensor network (WSN). Finally, the chapter concludes with the advantages and disadvantages of wireless networks.

10.2   Fundamentals of Wireless Networks

There is a wide variety of wireless technologies in use today: Wi-Fi, Bluetooth, ZigBee, NFC, WiMAX, LTE, HSPA, EV-DO, earlier 3G standards, satellite services, and more. Every one of the wireless technologies has its own set of constraints and limitations. Most wireless technologies operate on common principles and are subject to common performance criteria and constraints. For example, a wireless network contains base station gateways, access points, and wireless bridging relays. This section covers different aspects of wireless communication networks such as topology, electromagnetic interference, and the security measures that need to be taken to ensure a secure network.

Topology: The topology of a wireless network is the way network components (such as devices, routers, and gateways) are arranged.

Topology shows how wireless devices connect each other when there is no physical connection.

There are different wireless technologies designed for different topologies and use cases. The three common wireless topologies are star, mesh, and cluster tree.

1. *Star*: In this topology, individual wireless devices communicate directly with a central "hub," and devices communicate only with the hub. The distance between the device and the hub is limited to 30–100 meters. The star topology is also called point-to-point link. All packets between devices must go through hub, and for this reason the hub may become bottlenecked.

2. *Mesh:* The computing devices in a mesh topology can communicate with other devices or nodes in the network. A message can "hop" from node to node until it reaches the destination. This capability is called multi-hopping. When any node in a network can communicate with any other node, this is a multipoint-to-multipoint network (also known as an ad hoc or mesh network).

3. *Cluster tree*: This topology is a hybrid of star and mesh topologies. Wireless devices in a star (point-to-point) topology are clustered around routers or repeaters that communicate with each other and the gateway in a mesh (point-to-multipoint) topology.

Security: Security is an umbrella term encompassing the characteristics of authentication, integrity, privacy, and nonrepudiation. Although no computer network is completely secure, wired networks are generally more secure than wireless networks because wireless signals travel through the air and can easily be intercepted or eavesdropped. Wireless networks are vulnerable to security attacks in various protocol layers. While the security problems associated with wireless networking are serious, there are steps you can take to protect yourself. Encrypt the wireless traffic passing between your wireless access point device and your computers. Change your administrator password to one that is long. Be careful what you do online; someone could be monitoring your activity. If possible, for public access, use a VPN since VPNs encrypt connections at the sending and receiving ends.

Electromagnetic interference: Unlike wired networks, wireless networks are subject to electromagnetic interference (EMI) caused by other networks or devices nearby and can degrade the network.

Since devices can interfere with each other, every nation needs regulations that define the frequency ranges and transmission power for each technology that is permitted. There are two types of electromagnetic interference: (1) conducted interference and (2) radiated interference.

Example 10.1

Which wireless topology is right for you?

Solution

You choose the one whose characteristics best fit the main requirements of your application. A star topology is good for relatively simple applications where interruptions in communications would not prevent you from achieving desired results. A mesh topology is best if data reliability is crucial and you want to guard against data loss or signal degradation. A cluster tree may be used if you plan to apply different topologies to different parts of the network. No single topology works best in every application.

Conducted emissions are currents that are carried by metallic paths. Radiated electromagnetic interference consists essentially of the common mode noise, differential mode noise, and mixed mode noise signals. Radiated interference is a major issue in most wireless networks.

Implementation: The task of implementing a wireless network generally involves three steps:

Electromagnetic interference is the degradation in the performance of a device due to the fields making up the electromagnetic environment.

1. Establish the user requirements. Consider what the user wants to achieve with the network.
2. Establish the technical requirements: what attributes does the network need to possess in order to meet the requirements of the user.
3. Evaluate the available technologies, and see how they rank against the technical requirements.

It is needless to say that no single wireless technology can meet all requirements for customers in all environments. A number of wireless products can be mixed to build networks optimized for a given situation.

10.3 Wireless Local Area Network

Wireless LAN is perhaps the most commercially significant and most well developed of all wireless networks. Wireless LANs do not require cable as the traditional Ethernet-wired LANs. WLANs are designed to provide wireless access within a range of 100 meters. Thus, they are used mostly in homes, offices, schools, or laboratories. WLANs are based on IEEE 802.11 standards, which are marketed under the Wi-Fi brand name. Users of wireless LANs operate a number of devices, such as PCs, laptops, and PDAs.

Wireless LAN components include:

- *Wi-Fi hot spots:* Wi-Fi hot spots are available at many public places such as hotels, hospitals, and Starbucks coffee shops as wireless Internet access technology. When you are away from home or office, you can connect your laptops to the hot spots. Wi-Fi is erroneously regarded as "wireless fidelity," but this is not true. Wi-Fi is the name given by the Wi-Fi Alliance to the IEEE 803.11 WLAN standard. It provides wireless access to computing devices (such as laptop and tablet) using radio waves. It uses carrier sense multiple access (CSMA) to avoid transmission collisions. It has been promoted as the global WLAN standard.

- *Radio NIC:* Wireless network interface cards (NIC) are special network cards installed within wireless-enabled computers. NIC provides the interface between the computer device and wireless network infrastructure. It operates within the computer device and provides wireless connectivity. It is also known as a radio card, which implements the IEEE 802.11 standard. It picks up signals from the access point and converts them to signals the computer can understand. For computers, the NIC is usually network interface card that is installed in an expansion slot.

- *Access points:* An access point (or base station) in WLAN is a station that mainly transmits and receives data. This requires that antennas are built into most access points. The access point connects users to the network and can also connect the WLAN to a wired network. It has a limited range within which it can maintain a wireless connection with devices on the network. Performance usually suffers as distance from access point increases beyond the limits of a range of 50–90 meters.

- *Routers:* Routers are devices that literally route data around the network. A router transfers packets between networks. The router examines an arriving data and uses routing tables to determine the best path for the data to reach its destination. Routers use the software-configured network address to make decisions.

- *Repeaters:* A repeater simply regenerates a network signal to extend the range of the existing network infrastructure. It receives radio signals from an access point or another repeater; it retransmits the packets.

- *Print server.* A print server is used to connect printers to a network. This device allows everyone to use the same printer and saves cost.

- *Firewalls:* A firewall is a device, either hardware or software based, that controls access to

Fig. 10.2 A typical wireless local area network (WLAN). (Source: "10 different types of networks." https://www. networkstraining.com/ different-types-of-networks/)

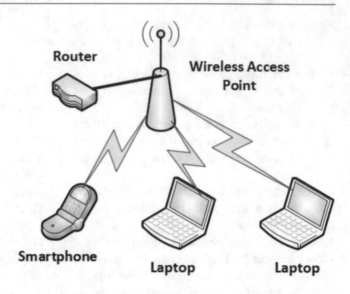

a company's network. A firewall does not perform differently on a wireless network that it does on a wired network. It is simply a device that separates two networks from each other. It is designed to protect data and resources from outside threat. For this reason, firewalls are typically placed at entry/exit points of a network so that it can deny or permit certain types of network traffic.

- *Antennas*: An antenna converts electrical signals to radio waves. A wireless network usually requires focused antennas (e.g., dish antenna) that can send a beam in a specific direction. For some applications, standard dipole antennas are sufficient. Nearly all access points, routers, and repeaters come with their own antennas. Smart or adaptive antennas are capable of rejecting interference and compensating for the multipath effects caused by signals reflecting off buildings and other structures. The performance requirements of future wireless systems cannot be met without the use of smart antennas.

These devices are used to create wireless LANs. A typical wireless LAN is shown in Fig. 10.2.

10.4   Wireless Metropolitan Area Network

A metropolitan area network (MAN) is a communication network that spreads over one or more neighboring cities. WMANs can connect LANs or WLANs without any cables by using microwave, radio wireless communication, or infrared laser which transmits data wirelessly. A wireless metropolitan area network (WMAN) is also known as a wireless local loop (WLL). WMAN is a new technology that will supplement well-known wired technologies such as synchronous optical network/synchronous digital hierarchy (SONET/SDH) and Gigabit Ethernet. It provides services to metropolitan or regional areas, either urban or rural, within a radius of 50 km.

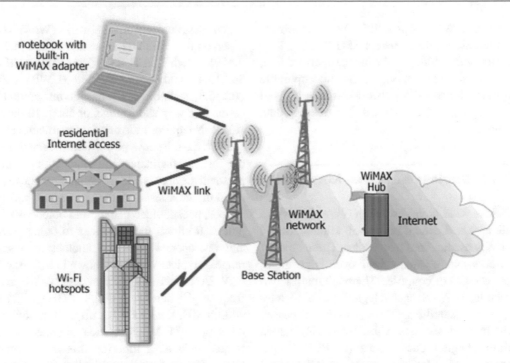

Fig. 10.3 An example of WiMAX. (Source: "Internet access guide." http://www.conniq.com/InternetAccess_WiMAX. htm)

A **wireless MAN** is intended to provide services over a geographical region of the size of a **metropolitan area**, approximately 50 kilometers or 30 miles

Standard WMANs are based on IEEE 802.16 which is often called WiMAX (Worldwide Interoperability for Microwave Access). WiMAX connects wireless LANs to form a WMAN without the need for expensive cabling. WiMAX is similar to Wi-Fi, but provides service to users within a 30-mile radius. It can provide wireless access to the Internet through a service that covers an entire metropolitan area. WLAN is generally owned and operated by a single organization, whereas a WMAN is usually used by individuals, organizations, and the public. Conceptually, WMANs can be used to connect WLANs/WPANs and provide access to data, voice, video, and multimedia services. WiMax is an international standard that provides a means of wirelessly connecting homes and businesses in metro area to the core communication networks. It can deliver triple play services (i.e., voice, video, and data) over microwave RF spectrum to stationary or moving users making broadband available anywhere.

WiMAX devices typically operate in the 2.5 GHz, 3.5 GHz, and 5.7 GHz frequency bands. A WiMAX link can transfer data at up to 70 Mbps. WiMAX employs time division multiple access (TDMA), which divides access to a given channel into multiple time slots. Each node transmits only in its assigned slot, thereby avoiding collisions. An example of WiMax is shown in Fig. 10.3.

Mobile WiMAX is based on using orthogonal frequency-division multiple access (OFDMA) as the physical layer technology to support multiple users in a scalable manner. WiMAX is profoundly changing the landscape of wireless broadband due to the variety of fundamentally different design options. It provides last mile connectivity to a backbone network such as the Internet. Scalability is a main challenge in such a network that covers a relatively large area.

Another WMAN is IEEE 802.20 or mobile broadband wireless access (MBWA), which enables worldwide deployment of affordable, ubiquitous, always-on, and interoperable multi-vendor MBWA networks. It is optimized for high-speed IP-based wireless data services.

10.5   Wireless Wide Area Network

Wireless wide area networks (WWAN) cover a wide area and extend beyond 50 kilometers and typically use licensed frequencies. These types of networks can be maintained over large areas, such as cities or countries. There are mainly two available WWAN technologies: cellular telephony and satellites. These will be discussed fully in later sections. Other WWANs include Cellular Digital Packet Data (CDPD), Global System for Mobile Communications (GSM), and Mobitex. WWAN technologies even offer Global Positioning System (GPS) that can locate a device anywhere in the world.

10.6   Wireless Personal Area Networks

This is a small-scale wireless network that connects a few devices in a single room.

Wireless personal area networks (WPANs) are based on the IEEE 802.15 standards and include the Bluetooth (IEEE 802.15.1), ZigBee (IEEE 802.15.4), and ultra-wideband (UWB) (IEEE 802.15.3) technologies. They permit communication in a very short range, of about 10 meters. A WPAN requires little or no infrastructure and is characterized by low power demands and a low bit rate. It facilitates communications among personal devices such as cellular phones, personal digital assistants (PDAs), pagers, personal stereos, pocket video games, and notebook computers. It allows these devices to communicate and interoperate with one another in a small range. A typical WPAN is shown in Fig. 10.4.

WPAN operates in the license-free radio frequency band of 2.4 GHz (2400–2483.5 MHz), with ranges up to 10 m and data rates up to 1 Mbps. It ties together closely related objects, a function that is fundamentally different from the objective of a wireless LAN. Thus, WPAN must coexist with other radio technologies, like WLAN, that operate the same frequency band.

WPAN is different from wireless LAN. First, WPANs target primarily the vast consumer market and are used for ease of connectivity of personal wearable or handheld devices. Second, WPAN is optimized for low complexity, low power, and low cost. It does not require access point like WLAN. Third, WPAN's close range

Fig. 10.4 A typical wireless PAN. (Source: Alagesan and Natarajan, 2020)

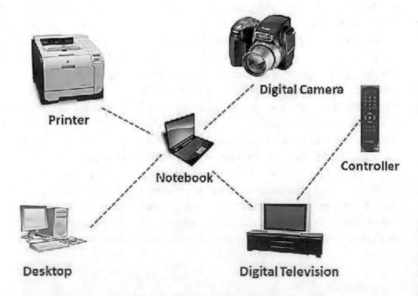

throughput of 1 Mbps does not compare with the 11 Mbps the IEEE 802.11 WLAN offers. It has small coverage, typically about 10 m, and connects only a limited number of devices.

WPAN and Bluetooth have been used interchangeably in many technical articles. Although the two technologies are similar, they should not be confused. Bluetooth is a far-field radio technology that is being promoted by the Bluetooth special interest group (SIP) with over 75 members. Bluetooth technology is for short-range, low-cost radio links between PCs, mobile phones, and other electronic devices. It provides a 10 m

lem with PAN is security. Since transferring of data across devices has become simple, a lot of work on encryption needs to be done.

10.7 Cellular Technologies

Perhaps no single development has done more for wireless technologies than cellular communications. It is one of the fastest growing and most demanding telecommunication applications. Areas of coverage are divided into small hexagonal radio coverage units known as *cells*.

A **cell** in the cellular network is the basic geographical unit in which the mobile users are located.

personal bubble that supports simultaneous transmission of both data and voice for multiple devices.

Some interesting applications of PAN include:

- Transfer of information between all the electronic devices a gadget-hungry person may carry. Such devices include pagers, cellular phones, personal digital assistants (PDAs), identification badges, and smart cards.
- Exchange of business cards by just shaking hands. The electronic card is transferred automatically from one card device via the body to the other person's card device.
- Office and domestic automation at home or at the office. One is able to control lighting, heating, and even locks with a gentle touch. These devices and appliances are programmed to communicate with each other with little or no human intervention.
- Telemedicine with intelligent sensors monitoring specific physiological signals such as EEG, ECG, and GSR and performing data acquisition. Such a collection of wearable medical sensors can communicate using PAN.

Other potential applications include sensor and automation needs, interactive toys, and location tracking for smart tags and badges. A major prob-

A cellular communications system employs a large number of low-power wireless transmitters to create the cells, which are base stations transmitting over small geographic areas that are represented as hexagons. Figure 10.5 illustrates communication coverage by spatial division to cells with base stations. Each cell size varies depending on the landscape and tele-density. Those stick towers one sees on hilltops with triangular structures at the top are cellular telephone sites. Each site typically covers an area of 15 miles across, depending on the local terrain. The cell sites are spaced over the area to provide a slightly overlapping blanket of coverage. The channel is made of two frequencies: one frequency (the forward link) for transmitting information to the base station and the other frequency (the reverse link) to receive from the base station. A typical cellular network is shown in Fig. 10.6.

10.7.1 Fundamental Features

Besides the idea of cells, the essential features of cellular systems include cell splitting, frequency reuse, handover, capacity, spectral efficiency, mobility, and roaming:

- *Cell splitting*: As a service area becomes full of users, the single area is split into smaller

Fig. 10.5 Cellular communication coverage by spatial division to cells with base stations. (Source: "Cellular communication network technologies." https://www.tnuda.org.il/en/physics-radiation/radio-frequency-rf-radiation/cellular-communication-network-technologies)

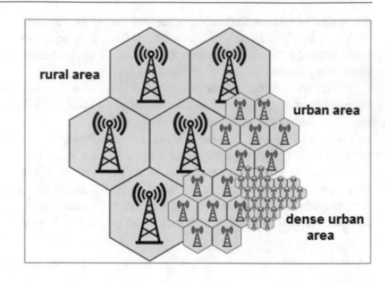

Fig. 10.6 A typical cellular network. (Source: https://www.123rf.com/photo_14506635_a--cellular-network-or-mobile-network-is-a-radio-network-distributed-over-land-areas-called-cells-eac.html)

ones. This way, urban regions with heavy traffic can be split into as many areas as necessary to provide acceptable service, while large cell can be used to cover remote rural regions. Cell splitting increases the capacity of the system.

the ability to reuse the same frequency many times.Cellular wireless networks can reuse radio channels in non-adjacent cells. Several frequency reuse patterns are in use in the cellular industry, each with its advantages and

Frequency reuse is the ability to reuse the same frequency (channel) several times.

- *Frequency reuse:* This is the core concept that defines the cellular system. The cellular telephone industry is faced with a dilemma: services are growing rapidly, and users are demanding more sophisticated call-handling features, but the amount of the electromagnetic (EM) spectrum allocation for cellular service is fixed. This dilemma is overcome by

disadvantages. Each cell is assigned three frequency groups. For example, the same frequencies are reused in cell designated 1, and adjacent locations do not reuse the same frequencies. A cluster is a group of cells; frequency reuse does not apply to clusters.

- *Handoff*: This is another fundamental feature of the cellular technology. When a call is in progress and the switch from one cell to another becomes necessary, a handoff takes place. Handoff is important because as a mobile user travels from one cell to another during a call, as adjacent cells do not use the same radio channels, a call must be either dropped or transferred from one channel to another. Dropping the call is not acceptable. Handoff was created to solve the problem. A number of algorithms are used to generate and process a handoff request and eventual handoff order. The need for handoff is determined by the quality of the signal, whether it is weak or strong. A handoff threshold is predefined. When the received signal level is weak and reaches the threshold, the system provides a stronger channel from an adjacent cell. This handoff process continues as the mobile

one access point to another access point in a neighboring cell.

- *Capacity*: This is the number of subscribers that can use the cellular system. Capacity expansion is required because cellular systems must serve more subscribers. It takes place through frequency reuse, cell splitting, planning, and redesigning of the system.
- *Spectral efficiency*: This is a performance measure of the efficient use of the frequency spectrum. It is the most desirable feature of a mobile communication system. It produces a measure of how efficiently space, frequency, and time are utilized. A cellular network (or mobile network) is a network distributed over land areas called cells, with each served by at least one fixed-location transceiver, known as a base station. The key challenge in mobile com-

Handing off is the process of transferring the mobile unit that has a call from a voice channel to another voice channel without interfering with the call.

moves from one cell to another as long as the mobile is in the coverage area.

Mobility: Mobility is one of the main benefits of wireless networks. It implies that a mobile

munications is handing off user communications from one local coverage area to the next.

Mobility enables the user to move around from one cell to another while maintainng the same call without service interruption.

user has the freedom to change location without losing connection.

- This is made possible by the built-in handoff mechanism that assigns a new frequency when the mobile moves to another cell. Due to several cellular operators within the same region using different equipment and a subscriber is only registered with one operator, some form of agreement is necessary to provide services to subscribers.
- *Roaming:* This is the process whereby a mobile moves out of its own territory and establishes a call from another territory. Cell-to-cell roaming takes place when a wireless client moves away from communicating with

10.7.2 Cellular Network

A typical cellular network consists of the following three major hardware components:

- *Cell site (base stations):* The cell site acts as the user-to-MTSO interface. It consists of a transmitter and two receivers per channel, an antenna, a controller, and data links to the cellular office. Up to 12 channels can operate within a cell depending on the coverage area.
- *Mobile Telephone Switching Office* (MTSO): This is the physical provider of connections between the base stations and the local exchange carrier. MTSO is also known as

mobile switching center (MSC) or digital multiplex switch-mobile telephone exchange (DMS-MTX) depending on the manufacturer. It manages and controls cell site equipment and connections. It supports multiple-access technologies such as AMPS, TDMA, CDMA, and CDPD. As a mobile moves from one cell to another, it must continually send messages to the MTSO to verify its location.

- *Cellular (mobile) handset*: This provides the interface between the user and the cellular system. It is essentially a transceiver with an antenna and is capable of tuning to all channels (666 frequencies) within a service area. It also has a handset and a number assignment module (NAM), which is a unique address given to each cellular phone.

- *Global System for Mobile Communications* (GSM): The GSM network is divided into three major systems: the switching system, the base station system, and the operation and support system. Its architecture was designed to handle international roaming. GSM is the most common standard and is used for a majority of cell phones. Since 1991, mobile radio has been dominated by the GSM system. GSM has become the most dominant system used in the world.

- *Personal communications service* (PCS): The GSM (Global System for Mobile Communications) digital network has pervaded Europe and Asia. A comparable technology known as personal communication systems (PCS) is making inroads in the USA. PCS is a relatively new concept that will expand the horizon of wireless communications beyond the limitations of current cellular systems to provide users with the means to communicate with anyone, anywhere, anytime. It refers to digital wireless communications and services operating at broadband (1900 MHz) or narrow-band (900 MHz) frequencies. Its goal is to provide integrated communications (such as voice, data, and video) between nomadic subscribers irrespective of time, location, and mobility patterns. It promises near-universal access to mobile telephony, messaging, paging, and data transfer. It

may be regarded as an extension of the cellular network to the 1900 MHz band.

- *Cellular generations:* Different cellular generations have been developed since early 1980s. First generation, 1G, was analog telecommunications standard introduced in the 1970s for voice communications with a data rate up to 2.4 kps. It used FM and FDMA and a bandwidth of 30 kHz. The major problems with 1G are poor voice quality, poor battery quality, and large phone size. There was no worldwide standard. The second generation, 2G, was based on digital technology and network infrastructure GSM (Global System for Mobile Communications), enabling text messages, and with a data speed of up to 64 Kbps. The digital cellular systems conform to at least three standards: GSM for Europe and international applications, one for the USA, and JDC for Japan. The third generation, 3G, was introduced in year 2000, with a data speed of up to 2 Mbps.

- The cellular systems use TDMA, CDMA, CSMA, and FDMA. It introduced high-speed Internet access. It used technologies such as W-CDMA and HSPA (high-speed packet access). It provided IP connectivity for real-time and non-real-time services. The development of 3G was mainly driven by demand for data services over the Internet. The fourth generation, 4G, may be regarded as the extension of 3G but with a faster Internet connection, more bandwidth, and a lower latency. It is capable of providing much higher data rates and any kind of service at anytime, anywhere. The expected rates will be in the range of 1Gb in the WLAN environment and 100 Mb in cellular networks. G4 technologies, such as WiMAX and LTE (Long-Term Evolution), claim to be about five times faster than 3G services. 5G or 5th generation wireless systems move us beyond networks design for mobile devices alone toward systems that connect different types of devices operating at high speeds. The 5G wireless technology is a multipurpose wireless network for mobile, fixed, and enterprise wireless applications. 5G will

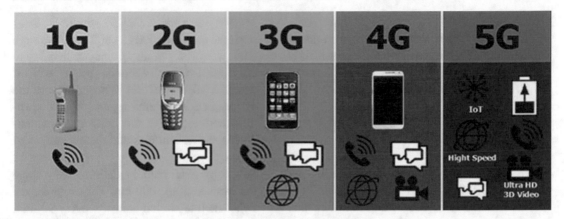

Fig. 10.7 Generations of cellular networks (Source: A. Nash, "Introduction about previous generations of mobile networks," https://mobiletrans.wondershare.com/5g/g-generations.html)

be a new mobile revolution as it is expected to provide gigabit-per-second data rates anytime, anywhere. In a 5G wireless network, every mobile phone will have an IPv6 address depending on the location and network being used. Although 5G technology promises a plethora of new applications, the technology is still in development stage. Figure 10.7 shows generations of wireless network.

10.8   Satellite Networks

Satellite (spacecraft) communications system consists of two parts, the space and ground segments, and is designed to implement go-anywhere-call-anywhere-from-anywhere concept.

The ultimate goal of communications systems is to provide pocket-sized, wireless devices that will accommodate voice and data services between any two locations throughout the globe. The new service that will achieve this goal ties together the past (satellite) and future (cellular) concepts to offer instant global connectivity—communications from anywhere to anywhere. When global coverage is needed, satellite networks are available. By extending communications to the remotest parts of the world, virtually everyone can be part of the global economy.

Satellites and UAVs (unpiloted aerial vehicles) have a set of unique characteristics when compared to ground-based communication nodes. Satellite communications are not a replacement of the existing terrestrial systems but rather an extension of wireless system. However, satellite communication has the following merits over terrestrial communications:

- *Coverage*: Satellites can cover a much large geographical area than the traditional ground-based system. They have the unique ability to cover the globe.
- *High bandwidth*: A Ka band (27–40 GHz) can deliver throughput of gigabits per second rate.
- *Low cost*: A satellite communications system is relatively inexpensive because there are no cable-laying costs and one satellite covers a large area.
- *Wireless communication*: Users can enjoy untethered mobile communication anywhere within the satellite coverage area.
- *Simple topology*: Satellite networks have simpler topology which results in more manageable network performance.
- *Broadcast/multicast*: Satellite are naturally attractive for broadcast/multicast applications.
- *Maintenance*: A typical satellite is designed to be unattended, requiring only minimal attention by customer personnel.
- *Immunity*: A satellite system will not suffer from disasters such as floods, fire, and earthquakes and will therefore be available as an emergency service should terrestrial services be knocked out.

Of course, satellites systems do have some disadvantages. These include propagation delay, dependency on a remote facility, and attenuation due to atmospheric particles (e.g., rain, snow) which can be severe at high frequencies.

10.8.1   Type of Satellites

There were only 150 satellites in orbit by September 1997. Currently there are over 2218 artificial satellites orbiting the earth. With this increasing trend in the number of satellites, there is a need to categorize them according to the height of their orbit and "footprint" or coverage on the earth's surface. They are classified as follows:

- *Geostationary earth orbit (GEO) satellites*: They are launched into a geostationary or geosynchronous orbit, which is 35,786 km above the equator. (Raising a satellite to such an attitude, however, required a rocket, so that the achievement of a GEO satellite did not take place until 1963.) A satellite is said to be in geostationary orbit when the space satellite is matched to the rotation of the earth at the equator. A GEO satellite can cover nearly one-third of the earth's surface, i.e., it takes three GEO satellites to provide global coverage. Due to their large coverage, GEO satellites are ideal for broadcasting and international communications. Examples of GEO satellite constellations are Spaceway designed by Boeing Satellite Systems and Astrolink by Lockheed Martin. There are at least three major objections to GEO satellites. First, there is a relatively long propagation delay (or latency) between the instant a signal is transmitted and when it returns to earth (about 240 milliseconds). Second, there is a lack of coverage at far northern and southern latitudes. Unfortunately, many of the European capitals, including London, Paris, Berlin, Warsaw, and Moscow, are north of this latitude. Third, both the mobile unit and the satellite of a GEO system require a high transmit power.

- *Middle earth orbit (MEO) satellites*: They orbit the earth at 5000–12,000 km. Although the lower orbit reduces propagation delay to only 60–140 milliseconds round trip, it takes 12 MEO satellites to cover most of the planet. MEO systems represent a compromise between LEO and GEO systems, balancing the advantages and disadvantages of each.

- *Low earth orbit (LEO) satellites*: They circle the earth at 500–3000 km. For example, the Echo satellite circled the earth every 90 min. To provide global coverage may require as many as 200 LEO satellites. Latency in a LEO system is comparable with terrestrial fiber optics, usually less than 30 millisecond round trip. LEO satellites are suitable for PCS. However, LEO systems have a shorter life space of 5–8 years (compared with 12–15 years for GEO systems) due to the increased amount of radiation in low earth orbit. An example is OrbComm designed by Orbital Corporation and consists of 36 satellites, each weighing 85 pounds.

A typical satellite network is shown in Fig. 10.8. The evolution from GEO to MEO and LEO satellites has resulted in a variety of global satellite systems. Several attractive services that can be offered by utilizing satellite technology include (1) personal communication service (PCS) in a global scale, (2) digital audio broadcasting (DAB), (3) environmental data collection and distribution, (4) remote sensing/earth observation, and (5) several military applications and satellite radio. Examples of satellite systems include Iridium, Globalstar, GPS, and ICO. The most popular is GPS (Global Positioning System), which is a US-owned utility that provides users with positioning, navigation, and timing services. This system consists of 24 satellites.

Every nation of the world has the right to access the satellite orbit, and no nation has a permanent right or priority to use any particular orbit location. Satellite services are classified into 17 categories: fixed, intersatellite, mobile, land mobile, maritime mobile, aeronautical mobile,

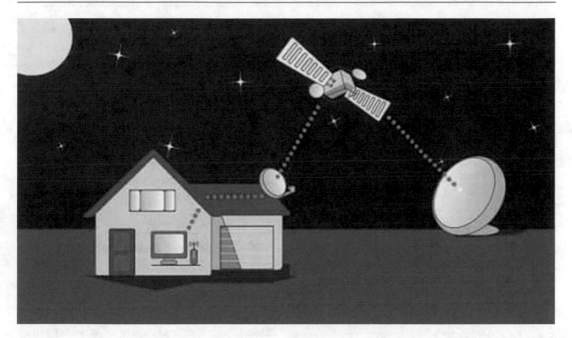

Fig. 10.8 A typical satellite network. (Source: Rajiv, 2022)

broadcasting, earth exploration, space research, meteorological, space operation, amateur, radio-determination, radio navigation, maritime radio navigation, and standard frequency and time signal. The International Telecommunication Union (ITU) is responsible for allocating frequencies to satellite services. The Ku band (12–18 GHz) is presently used for broadcasting services and also for certain fixed satellite services. The C band (4–8 GHz) is exclusively for fixed satellite services, and no broadcasting is allowed. The L band (1–2 GHz) is employed by mobile satellite services and navigation systems.

10.8.2   Satellite Components

Every satellite communication involves the transmission of information from a ground station to the satellite (the uplink), followed by a retransmission of the information from the satellite back to the earth (the downlink). Hence the satellite must typically have a receiver antenna, a receiver, a transmitter antenna, a transmitter, some mechanism for connecting the uplink with the downlink, and a power source to run the electronic system. These components are explained as follows:

- *Transmitters*: The amount of power required by a satellite transmitter to send out depends on whether it is GEO or LEO satellite. The GEO satellite is about 100 times away as the LEO satellite. Thus, GEO would need 10,000 times as much power as LEO satellite. Fortunately, other parameters can be adjusted to reduce this amount of power.
- *Antennas*: The antennas dominate the appearance of a communication satellite. The antenna design is one of the more difficult and challenging parts of a communication satellite project. The antenna geometry is constrained

physically by the design and the satellite topology.

Power generation: The satellite must generate

Transponders: These are the devices each satellite must carry. They receive radio signals at one frequency, amplify them, and convert them to

A **wireless sensor network** (WSN) usually consists of a large number (hundreds or thousands) of sensor nodes deployed over a geographical region.

Example 10.2

Describe Global Positioning System (GPS).

Solution: The Global Positioning System (GPS) is a common application of satellite communication. GPS is a satellite-based navigation system made up of a network of 24 satellites placed into orbit by the US Department of Defense. GPS was originally designed for military use, but in the 1980s, the government made the system available for civilian use.

The 24 satellites that make up the GPS space segment are orbiting the earth about 12,000 miles above us. These satellites travel at speeds of roughly 7000 miles an hour. GPS satellites transmit two low-power radio signals. The signals travel by line of sight, meaning they will pass through clouds, glass, and plastic but will not go through most solid objects such as buildings and mountains.

A GPS receiver calculates its position by precisely timing the signals sent by GPS satellites. The receiver uses the messages it receives to determine the transit time of each message and computes the distance to each satellite. These distances along with the satellites' locations are used to compute the position of the receiver.

all of its own power. The power is often generated by large solar cells, which convert sunlight into electricity. Since there a limit to how large the solar panel can be, there is also a practical limit to the amount of power which can be generated.

another for transmission. For example, a GEO satellite may have 24 transponders with each assigned a pair of frequencies (uplink and downlink frequencies).

10.9   Wireless Sensor Networks

Wireless sensor networks (WSNs) are the newest of wireless networks. Recent advances in wireless communication and embedded systems have caused the rapid deployment of wireless sensor networks in all sectors and has started to find its place in industrial environment. WSN is a technology with promising future and a wide range of applications.

A wireless sensor device consists of a sensor that measures a physical phenomenon. The wireless sensor nodes are compact, light-weighted, and battery-powered devices that can be used in virtually any environment. Each node functions as a sensor node as well as a router node. The sensor nodes monitor physical or environmental conditions such as temperature/ heat, humidity, sound, vibration, pressure, light, object motion, pollutants, presence of certain objects, noise level or characteristics of an object such as weight, size, speed, direction, and its latest position. The sensor node is made up of four components: a power unit, a transceiver unit, a sensing unit, and a processing unit. The node may also have some application-dependent components. The sensor market has been segmented into temperature sensor networks, pressure sensor networks, level sensor networks, flow sensor networks, humidity sensor networks, motion and position sensor networks, gas sensor networks, light sensor networks, chemical sensor networks, and others.

WSNs belong to the general family of sensor networks that employ distributed sensors to collect information on entities of interest. In general,

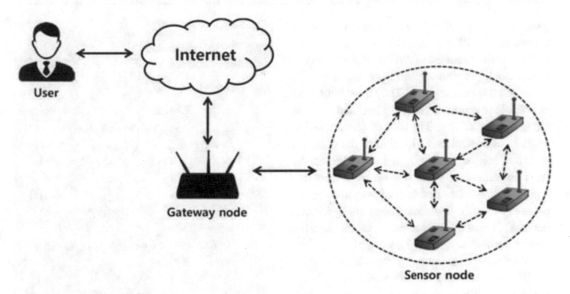

Fig. 10.9 A typical wireless sensor network. (Source: Yu and Park, 2020)

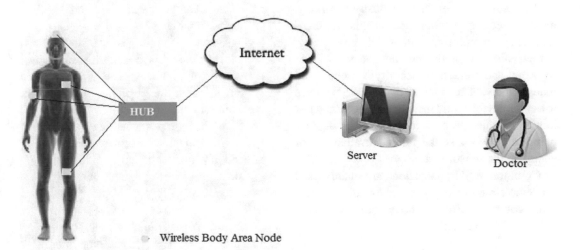

Fig. 10.10 A typical WBAN. (Source: "PhD in wireless body area networks." http://phdinfo.org/PhD_in_ WirelessBodyAreaNetwork.html)

there may be both sensing and non-sensing nodes in a WSN; i.e., all sensors are nodes but not all nodes are sensors. A sensor has four operating modes: transmission, reception, idle listening, and sleep. Collision occurs when two or more nodes transmit at the same time. A sensor node is designed to use an operating system (OS). TinyOS (developed at UC Berkeley) is perhaps the first operating system specifically designed for WSNs. Most WSNs are application specific and are designed to meet the challenges for that

specific application. Applications of WSNs typically involve some kind of monitoring, tracking, or controlling. Network topologies in a WSN may range from star to a full mesh topology. The way the sensors are connected determines whether the sensors can be safely, securely, and cost-effectively deployed in the typical harsh industrial environments. One of the objectives of WSNs is to minimize the running cost, and therefore star and tree topologies are used because they do not require any routing table to send the packets. A typical wireless sensor network is shown in Fig. 10.9.

The main features of WSNs include robustness, scalability, self-organization, extended lifetime, and low cost and size. They have limitations in complexity, power consumption, and communication capabilities. The cost of sensor components is of critical consideration in the design of practical sensor networks. It is often economically expedient to discard a sensor rather than recharge a sensor. For this reason, the battery power is usually a scare component in wireless devices. It causes vulnerability of the wireless sensor networks.

Wireless sensor networks were initially motivated by the military for battlefield surveillance, but they constitute the platform of a broad range of applications in agriculture, homes, offices, governments, universities, industry, military, national security, surveillance, healthcare, environmental monitoring, industrial monitoring, natural disaster prediction, healthcare, vehicle tracking and detection, security systems, and many others. The state-of-the-art WSNs have lower deployment and maintenance costs and last longer. A major problem in the practical deployment of a WSN is the limited availability of energy at the sensor nodes.

Common WSN applications are mainly used for remote monitoring, home automation, industrial control, wireless body area networks (WBANs), and tracking.

- *Monitoring:* Wireless sensor networks facilitate the monitoring and controlling of factories, offices, homes, vehicles, cities, and plants. They can monitor physical and environmental conditions. The monitoring system may be based on low-power ZigBee wireless communication technology. It is usually made up of sensor node, the sink node, and transmission networks. The monitoring may involve physical parameters such as temperature, humidity, and light. For example, a major pharmaceutical manufacturer may decide to connect all of its R&D equipment to the company's control systems for 24/7/365 monitoring. Monitoring may also include water/

Example 10.3

Consider the following applications of WSNs.

(a) In the chemical industries, manufacturing wastes are stored and processed in a particular location at a manufacturing or processing site. By implementing efficient monitoring sensors in and around the chemical industrial environment, the hazards of chemical wastes can be minimized.

(b) A stream biologist uses data from upstream flow sensors to increase the collection rate of downstream samplers so that the impact of a full flood cycle on stream chemistry can be assessed at the peak of the flood.

(c) Posts, removing distance between the researcher and observational sites.

(d) A technician replaces a failing oxygen sensor within hours of the failure of its predecessor, thus avoiding the loss of data for a month or more.

(e) A graduate student counting plants in a permanent plot uses an online key to aid in plant identification and confirms difficult identifications through a videoconference link to an advisor.

Each of these examples illustrates how wireless sensor networks are emerging as revolutionary tools for studying complex real-world systems.

petroleum pipeline monitoring, security monitoring, environmental monitoring, meteorological monitoring, weather forecasting, and pollution monitoring.

- *Tracking:* The sensors nodes were deployed in such a way that more than one node would be triggered if an intruder enters the area covered by the WSN. This makes them suitable for tracking applications. These include tracking objects, animals, humans, and vehicles. For

Example 10.4

Discuss the differences between WBAN and WPAN.

Solution:

The development of WBAN technology started around 1995 with the idea of using wireless personal area network (WPAN) technologies to implement communications on, near, and around the human body. Some years later, the term "BAN" came to refer to systems where communication is entirely within, on, and in the immediate proximity of a human body. Thus, WBAN is not exactly as WPAN. While a WBAN (such as a wearable computer) operates within the range 1–2 m, a WPAN (such as Bluetooth) operates in the range of 10 m. However, WBAN complements WPANs. WBANs target diverse applications including healthcare, athletic training, workplace safety, consumer electronics, secure authentication, and safeguarding of uniformed personnel. Due to these diverse application of WBANs, it seems WBANs are more popular than WPAN.

example, one can track the development of an acidic chemical plume applied to the monitoring environment through a point source.

- *Wireless body area networks* (WBANs): These are a special purpose WSNs that can provide ubiquitous real-time monitoring of human physiology. A WBAN is a network formed by low-power devices that are located on, in, or around the human body. In these networks, various sensors are attached on clothing, the body, or even implanted under the skin. For example, the sensors can measure the heartbeat, body temperature, or electrocardiogram. WBANs can be used in healthcare, fitness, gaming and entertainment, military, etc. For example, WBANs can be used for proactive

monitoring and treatment of a personal health. A typical WBAN is shown in Fig. 10.10.

10.10 Advantages and Disadvantages

The advantages or benefits of wireless networks include:

- *Convenience:* You can access online resources from any location anytime. Through ubiquitous connectivity, you can check email and browse or talk on the run regardless of location, time, or circumstance.
- *Mobility:* This is a major advantage offered by wireless networks. The wireless network frees the user from the cord. It allows you to move or roam and still be connected. Wireless and mobile networks are quickly becoming the networks of choice due to the flexibility and freedom they offer.
- *Easy installation:* Wireless networks can easily be set up and dissembled. Installation is quick, cost-effective, and adaptive. It is easy to add other components (such as printers, VOIP) to the network. A wireless network can be planned and implemented in days rather than months.
- *Expandability:* It is easy to expand the capacity of a wireless network to serve the needs of a growing business.
- *Operating cost:* Wireless networks cost less to operate than wired networks since they require no cables.
- *Productivity:* Wireless networks are a powerful tool for boosting productivity and encouraging information sharing.

The disadvantages or drawbacks of wireless networks include:

- *Security:* This is the biggest disadvantage of using a wireless network. Wireless networks have their own problems such as unreliability and security. If not installed and maintained properly, wireless network can be a security risk due to eavesdropping. Although encryption techniques and password protection can be used for security, some of the encryption techniques can be easily compromised.

- *Limited bandwidth:* Wireless networks have limited bandwidth. For example, they cannot support video teleconferencing. The maximum speed of IEEE 802.11n standard network is 600 Mbps, which is only almost half the speed of a wired network.
- *Limited range:* The typical range of a medium-range wireless network is up to around 100 meters. This may be good enough for a home or a small office, but insufficient for larger structures.
- *Interference:* Wireless networks are subject to jamming, electromagnetic interference, and fluctuations in available bandwidth. Wired LANs offer superior performance.
- *Cost:* Wireless components often cost more than the equivalent wired Ethernet products. For example, wireless access points may cost three or four times as much as Ethernet cable adapters.

Summary

1. Wireless network is the technology that provides seamless access to information anywhere, anyplace, and anytime without wires or cables.
2. Wireless LAN allows laptops, PCs, and users to link through radio waves or infrared links, eliminating the need for restrictive cables. Wi-Fi is a matured and popular standard for WLAN due to its use in hot spots (hotel, airport, etc.).
3. A wireless MAN is a new technology that provides services to metropolitan or regional areas within a radius of 50 km. WiMAX is a standard WMAN based on IEEE 802.16 and can provide wireless access to the Internet through a service that covers an entire metropolitan area.
4. Wireless WAN covers a wide area and extends beyond 50 kilometers. There are mainly two available WWAN technologies: cellular telephony and satellites.
5. Wireless PAN is a near-field intrabody communication. It is a wireless connection between PCs, peripherals, and portables that will let the devices share information without having to make a physical connection.
6. A cellular network is based on dividing a geographic area into cells, with each cell having its own transmitter-receiver (antenna) under the control of a base station. It operates on the principles of cell, frequency reuse, and handoff.
7. Satellite networks are used when global coverage is needed. Each satellite is equipped with various transponders consisting of a transceiver and an antenna.
8. A wireless sensor network (WSN) consists of a large number of sensor nodes deployed over a geographical region with the sensor nodes monitoring physical or environmental conditions.

Review Questions

10.1 Which of these networks is not wireless?
(a) Ethernet   (b) IEEE 801.11   (c) Cellular network   (d) Satellite

10.2 Which of the following networks covers the widest range?
(a) WLAN   (b) WMAN   (c) WWAN   (d) WPAN

10.3 Which of the following networks covers a home or building?
(a) Wi-Fi   (b) WiMAX   (b) 3G and 4G   (d) Satellite

10.4 The following are needed in a typical wireless LAN except:
(a) Antenna   (b) Base station   (c) Router   (d) Gateway

10.5 Which of the following is not a characteristic of a cellular network?
(a) High bandwidth   (b) Mobility   (c) Frequency reuse   (d) Cell splitting

10.6 Which of these is not a cellular technology:
(a) GSM   (b) PCS   (c) G4   (d) GEO

10.7. Which of the following bands is not used for satellite communication?
(a) MF   (b) Ku   (c) L   (d) C

10.8 The following are applications of wireless sensor networks except:

(a)   Tracking   (b)   Monitoring   (c) Automation   (d) Wireless LAN

10.9 The following are advantages of wireless networks except:

(a)   Mobility   (b)   Comfort   (c) Productivity   (d) Expandability

10.10 The following are disadvantages of wireless network except:

(a) Limited   coverage   (b) Limited   bandwidth   (c)   Limited   range   (d) Electromagnetic interference

Answer: 10.1 (a), 10.2 (c), 10.3 (a), 10.4 (d), 10.5 (a), 10.6 (d), 10.7 (a), 10.8 (d), 10.9 (b), 10.10 (a)

Problems

10.1 Explain   the   concept   of   wireless communication.

10.2 Mention five examples of wireless networks.

10.3 What is a radio NIC?

10.4 Discuss the following terms: antenna, base station, router, and repeater.

10.5 Discuss the star and mesh wireless topologies.

10.6 What is electromagnetic interference? Mention its two types.

10.7 What kinds of security measures can you take to protect yourself?

10.8 Describe WiMAX. How is it different from Wi-Fi?

10.9 Do some research on Long-Term Evolution (LTE) as a wireless MAN.

10.10 Describe wireless personal networks.

10.11 What is meant by Bluetooth?

10.12 In what ways is WPAN different from WLAN?

10.13 What is meant by a cell?

10.14 What is mobility?

10.15 What do we mean by roaming?

10.16 Explain the following terms: cell splitting, frequency reuse, and handoff in a cellular communications system.

10.17 What is the function of the base station in cellular network?

10.18 What are the functions of MTSO?

10.19 Discuss cellular mobile networks 1G, 2G, and 3G.

10.20 What is the difference between 3G and 4G?

10.21 What is GSM?

10.22 Describe PCS.

10.23 What is a satellite used for?

10.24 Discuss some of the advantages of satellite communication   over   terrestrial communications.

10.25 Mention some disadvantages of satellite networks.

10.26 What are the differences between GEO and LEO satellites.

10.27 Explain the following terms used in satellite communications: antenna, transmitter, and receiver

10.28 Mention some of the services satellite technologies provide.

10.29 Describe a sensor node.

10.30 What is meant by a wireless sensor network?

10.31 Mention some areas where WSN is being applied.

10.32 Discuss the pros and cons of wireless sensor networks.

10.33 Describe two ways in which WSNs are used.

10.34 What is the major issue in deploying a WSN?

10.35 What are the advantages of wireless networks?

10.36 Discuss three disadvantages of wireless networks.

Network Security

11

Know yourself and know your enemy, and you will not be defeated.

Sunzhi

Abstract

Network security is very much needed by users and more so by businesses. The study of this very topic is important to learning how security is implemented and maintained in any network system. This chapter covers the important aspects of network security. It also characterizes what malware, firewall, encryption, digital signatures, intrusion detection and prevention, and cybersecurity issues are and how they relate to network security.

Keywords

Network security · Malware · Firewall · Encryption · Intrusion detection · Intrusion prevention · Cybersecurity

11.1 Introduction

The increasing need for computer networking worldwide has created an urgent need for network security. The need is also important in the age of different abuses by hackers and others that want to cause harm to the information that is transmitted or stored in these computer networks. Therefore network security is very much needed by users and more so by businesses. In this chapter, the important aspects of network security are discussed. The chapter characterizes what malware, firewall, encryption, digital signatures, intrusion detection and prevention, and cybersecurity issues are and how they relate to network security.

11.2 Malware

Malware can be defined as software whose intent is nothing but to be malicious to systems such as computer system networks.

Malwares are always very harmful, destructive, intrusive, and hostile, with the goal of trying to destroy, disable, invade, or damage the computer network or any of the operational networks that can be used to communicate information from one system to the other.

The different types of malware include spyware, computer viruses, Trojan horses, and worms. While malware cannot damage the physical computer system network itself, it has the capability to hijack, delete, encrypt, and steal the information that is in the hardware system. It also has the capability to hide in the computer network and spy without the permission of the owner or user on any activity that is being performed with the network hardware.

11.2.1 The Function of a Malware

There are different types of malware, and each one has its own functional characteristics that help to distinguish what kind of malware it is, especially in the way the malware proceeds to cause damage to the network. Some of the common types of the malware are as follows:

- *Trojans:* The Trojan malware makes you believe that it is the real software in the system network. It has a discrete behavior and the tendency to make it possible for other malwares to make their way into the system network.
- *Spyware:* Just as the name sounds, the spyware hides inside the system network spying on all the activities going on in the system's network. A few examples of such activities the spyware can monitor and spy on are when you enter your login and credit card information and the websites you go to in the Internet.
- *Virus:* Viruses can take the form of an executable file. In its characteristic mode, it attaches itself to a clean file and then proceeds to infect and damage the rest of the clean files in the system network. Viruses have the functional capability to spread very fast and cause lots of damage if not completely corrupting or deleting files they find within reach.
- *Worms:* Worms have the ability to infect the entire networks both locally and across the Internet. The worms infect each networked system consecutively.
- *Botnets:* When you have networks of infected computer and they work together in a controlling manner by an attacker, they are called botnets or botnet malwares.
- *Adware:* This is the kind of malicious malware that keeps on popping up all kinds of annoying advertisements on the screen and makes the security of system network terrible to manage. It is very harmful and does slow work using the system network.
- *Ransomware:* This is a very dangerous malware that attempts to destroy and erase all that you have in the network and locks down the computer network. The worse is that it demands some ransom to be paid by the owner less it goes on the destroying and erasing mode hence the name ransomware. It is also called scareware.

11.2.2 Malware Removal Solutions

Malwares are characteristically different. They therefore require different ways of removing them once they have infected the network. However, installing and using anti-malware software is one of the best ways to take care of malwares. These anti-malwares are also called antivirus software. The other way that can be used in the removal of malwares that have infected the network is to carefully avoid suspicious emails, websites, and links. Some of the anti-malware removal software tools are:

- *TotalAv:* This anti-malware protects all network system devices such as Windows, Mac, Android, and iOS. It protects against malware, adware, and spyware.
- *Scanguard:* This is a malware-destroying software. It scans and removes viruses, spyware, and malware.
- *Norton:* The Norton anti-malware removes malware, spyware, and adware, blocks phishing websites, and provides Internet security and web protection.
- *PC Protect:* It protects against all threats including optimizing PCs and provides web security.
- *Malwarebytes:* It is an anti-malware software tool that removes malware, spyware, and adware, provides web protection, and has VPN Internet security.
- *McAfee:* This anti-malware removes malware, spyware, and adware, provides web protection, and has VPN Internet security.
- *Avira:* The Avira anti-malware software tool removes malware, spyware, and adware, blocks phishing websites, has VPN Internet security, and provides firewall and web protection including browser management.

- *BullGuard:* The BullGuard anti-malware software tool removes malware, spyware, and adware, blocks phishing websites, has VPN Internet security, and provides firewall and web protection including browser management.
- *Avast:* The Avast anti-malware software tool has the capability to instantly identify harmful threats. It speeds up overall performance. It deletes redundant files and secures Internet connection.
- *Bitdefender:* The Bitdefender anti-malware software tool removes malware, spyware, and adware, blocks phishing websites, has VPN Internet security, and provides firewall and web protection including browser management.

operates by monitoring the fields in the header of each packet. When it is combined with circuit and application gateways, it becomes what can be called a dynamic packet filtering process. The circuit gateway monitors the Transmission Control Protocol (TCP) handshaking between packets to determine whether a requested session is legitimate while the application level gateway is implemented through a proxy server, which acts as an intermediary between a client and a server. The packet filter is used to filter incoming or outgoing interfaces such as ingress filtering of spoofed Internet Protocol (IP) addresses and egress filtering. It permits or denies certain services and requires intimate knowledge of the TCP and the User Datagram Protocol (UDP) port utilization on a number of operating systems. Figure 11.2 shows the strategic positioning of a firewall between the local area network (LAN) and the wide area network (WAN). This helps to

11.3   Firewall

Any network security system that has the capability of monitoring both incoming and outgoing network traffic based on a well-defined security rules and uses information to determine whether to block or allow a specific traffic to the network system is known as a firewall security system.

11.3.1 The Firewall Protection

The illustration of how a firewall can protect an organization from unwanted traffic from an Internet network system is shown in Fig. 11.1. The packet filter is one of the firewall mechanisms that are used to protect the network of an organization. It is placed in the router that connects the organization's network to the Internet. It

prevent unwanted information from the LAN to the WAN providing the much needed network security for WAN.

11.3.2 Benefits of Having a Firewall

The following are some of the benefits of having a firewall for network security.

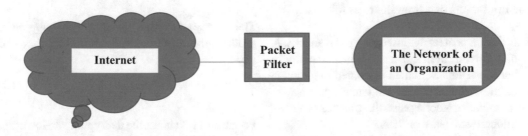

Fig. 11.1 An illustration of a firewall using packet filter that protects the organization's network against unwanted traffic from the Internet

Fig. 11.2 An
illustration of a firewall
between LAN and WAN

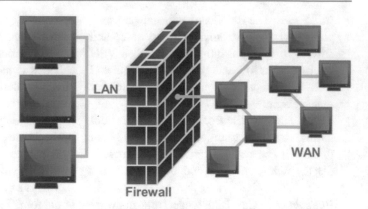

- It offers a central point of contact for information delivery service to customers.
- It is the perfect point to audit or log Internet usage.
- A firewall is a logical place to deploy a network address translator (NAT) that can help alleviate the address space shortage and eliminate the need to renumber when an organization changes Internet service providers (ISPs).
- If an organization does not have a firewall, each host system on the private network is exposed to attacks from other hosts on the Internet.
- In general, firewalls offer a convenient point where Internet network security can be monitored and alarms generated.

11.3.3 The Limitations of a Firewall

The limitations of a firewall are as follows:

- It cannot protect against data-driven attacks. A data-driven attack occurs when seemingly harmless data is mailed or copied to an internal host and is executed to launch an attack.
- It cannot protect against the transfer of virus-infected software or files.
- It cannot protect against the types of threats posed by traitors or unwitting users.
- It creates a single point of failure.

- It cannot protect against attacks that do not go through the firewall.

11.4 Encryption

Encryption is the mechanism used to ensure that the content of the information from point A (source) to point B (destination) still retains its original content and integrity.

There are several methods by which encryption can be implemented. Generally as an example, the sender and the receiver must both have a copy of the encryption key, and it is kept secret. The sender uses the key to produce encrypted information that is transmitted across the network. The receiver uses the key to decode the encrypted information. That is, the *encrypt* function used by the sender takes two arguments, namely, a key, K, and the information to be encrypted, I. The function produces an encrypted, E, version of the information I, as shown in Eq. 11.1.

$$E = encrypt\left(K,I\right) \qquad (11.1)$$

The *decrypt* function reverses the mapping to produce the original information I as shown in Eq. 11.2.

$$I = decrypt\left(K,E\right) \qquad (11.2)$$

Equation 11.3 shows the decrypted version of the encrypted information I:

$$I = decrypt\left(K,encrypt\left(K,I\right)\right) \qquad (11.3)$$

Example 11.1 Using additive cipher, let the alpha numeric assignments be a = 1 to z = 26. Let the key (*K*) = 14 and the information (*I*) to be encrypted (*E*) be "hello." Therefore, let:

Plaintext: h → 08	Encryption: (08 + 14) mod 25	Cipher text: 22 → V
Plaintext: e → 05	Encryption: (05 + 14) mod 25	Cipher text: 19 → S
Plaintext: l → 12	Encryption: (12 + 14) mod 25	Cipher text: 01 → A
Plaintext: l → 12	Encryption: (12 + 14) mod 25	Cipher text: 01 → A
Plaintext: o → 15	Encryption: (15 + 14) mod 25	Cipher text: 04 → D

The cipher text is therefore "VSAAD."

11.4.1 Public Key Encryption

Figure 11.3 shows the simplified model of the conventional encryption. In the encryption implementation, the key must be kept secret so that the security of the network is not compromised. In a specific example, each of the users is assigned a pair of keys. One of those keys given to the user is called the *private key*, and it must be kept secret, while the other, called the *public key*, is made public along with name of the user so that everyone knows the value of the key as shown in Fig. 11.4. The mathematical relationship as shown in Eqs. 11.4 and 11.5 of the encryption function has the property that the information encrypted with the public key cannot be easily decrypted except with the private key and the information encrypted with the private key cannot be decrypted except with the public key.

The relationships between the encryption and the decryption as with the two keys can be expressed mathematically as follows. Let *I* denote the information, **pub-ul** denote the *I*'s public key, and **prv-ul** denote the use *I*'s private key as shown in Eqs. 11.4 and 11.5, respectively. Therefore,

$$I = decrypt\left(pub-ul, encrypt\left(prv-ul, I\right)\right) \tag{11.4}$$

and

$$I = decrypt\left(prv-ul, encrypt\left(pub-ul, I\right)\right) \tag{11.5}$$

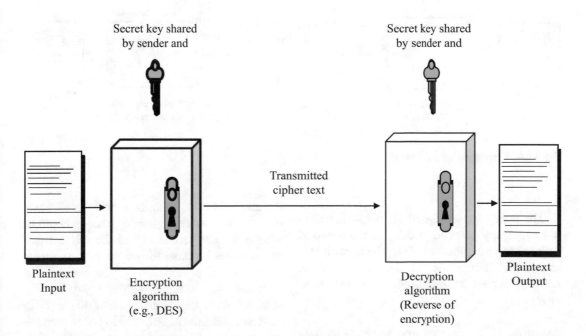

Fig. 11.3 The model of the conventional encryption: a simplified version

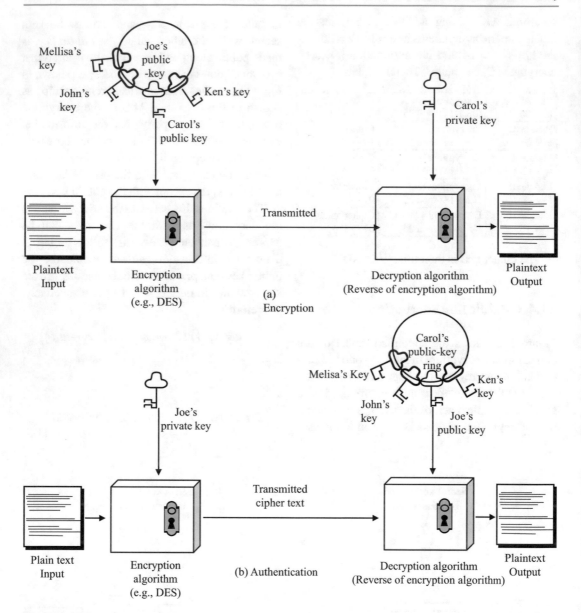

Fig. 11.4 Cryptography: public key

The public knowledge of the public key does not cause any harm to security of the network. This is because if the user has a private key that is secret, the information and the security of the network are not compromised in any way shape or form.

Example 11.2 Let each user generate its own pair of keys for the encryption and decryption of the information to be transmitted. Then each of the users (Mellisa, John, Ken, and Carol) place two keys in a public register or where it could be accessible. The second key of the original pair that each has becomes the private key. As shown in Fig. 11.4a, b, respectively, you can see that each user has a set of public keys that was obtained from the others. Should Joe wish to send a confidential information to Carol, then Joe will encrypt the information with Carol's public key. When Carol receives the information, she will decrypt it with her private key. It should be noted that no other recipient can decrypt the information because it is only Carol that knows Carol's private key. The same process is what all of the other participants will follow in their communication processes using their public and private keys as shown in Fig. 11.4a, b, encryption and authentication processes, respectively.

11.4.2   Types of Encryptions

There are basically three types of encryptions: transparent encryption, semi-transparent encryption, and manual encryption. In the transparent type, encryption is performed at low-level during all operations permanently. It is difficult to implement correctly and generally it does not work well with networking. It is easy to use and most secure.

In the semi-transparent encryption also known as "on-the-fly encryption," it does not operate permanently; however, it is before or after access. It has the potential capability to cause degradation of the computer's efficiency. When the data to be encrypted are too great, it can possibly cause a loss of data.

In the case of the manual encryption, the process is completely initiated by the user. It demands the user's active participation. Even though there is element of risk involved, it is reliable.

11.4.3   Encryptions and Decryptions

In the *symmetric/private key encryption* case, a single number key is used to encode and decode the data. The sender and receiver both must know the key. The Data Encryption Standard (DES) is the most widely used standard for symmetric encryption. This is because each sender and receiver would require a different key. This type of encryption is basically used by government entities. It is rarely used for e-commerce transactions over the Internet. It requires a secure way to get the key to both parties.

In the *asymmetric/public key encryption* case, there is no secret in encryption; there is secret in

decryption. Its primary benefit is that it allows people who have no preexisting security arrangement to exchange information securely. The need for sender and receiver both sharing the secret keys through some secure channel is eliminated. All of the communications involve only public keys, and no private key is ever transmitted or shared. Some of the examples of public key cryptosystems were invented by Ron Rivest, Adi Shamir, and Leonard Adleman, named (RSA); by Diffie and Hellman, named Diffie-Hellman; by David Kravitz, named the Digital Signature Algorithm (DSA); and by Taher Elgamal, named Elgamal. In addition, we have the Pretty Good Privacy (PGA) which is fairly popular and inexpensive. As a result of the fact that the conventional cryptography was once the only available means for relaying secret information, the expense of secure channels and key distribution relegated its use only to those who could afford it, such as governments and large banks. Public key encryption is the technological revolution that provides strong cryptography to the public.

11.4.4   Digital Certificates and Digital Signatures

A digital certificate is a unique identifier assigned to a user by a certification authority to verify the identity of the user. A certification authority vendor becomes the private company that certifies the user whether the person is who the individual claims to be. They work together with the credit card verification companies or other financial institutions in order to verify the identity of the certificate's requesters. The digital certificate received by the user includes a copy of its public key. This digital certificate's owner makes its public key available to anyone wanting to send encrypted documents to the certificate's owner.

The two main functions of the digital signature are to identify the signer and the formal approval of the document to be transmitted. Asymmetric encryption is used to create digital signatures. It is used on the Internet to authenticate both users and vendors. Digital signature is an encrypted attachment added to the electronic message to verify the sender's identity. Instead of encrypting information using someone else's public key, you encrypt it with your private key. If the information can be decrypted with your public key, then it must have originated with you.

11.4.5 Encryption Standards

The Advanced Encryption Standard (AES) is one of the encryption standards adopted by the U.S. government. It comprises three block ciphers namely, AES-128, AES-192, and AES-256. Each AES cipher has a 128-bit block size and key sizes of 128, 192, and 256 bits.

The Data Encryption Standard (DES) is one of the other encryption standards. It is a symmetric key block cipher adopted by the National Institute of Standards and Technology (NIST) in the year 1977. It is based on the Feistel structure where

The key size of DES is 56 bit which is comparatively smaller than AES which has 128-,192-, or 256-bit secret key. The rounds in DES include Expansion Permutation, Xor, S-box, P-box, Xor, and Swap. On the other hand, rounds in AES include SubBytes, ShiftRows, MixColumns, and AddRoundKeys. DES is less secure than AES because of the small key size. The AES is comparatively faster than DES.

11.5   Intrusion Detection and Prevention Systems

In the network security world, there is always an attempt to breach the security of the network by individuals that want to cause harm to the network. Intrusion detection and prevention systems become very essential in such networks to be able to maintain the integrity of the information that goes through such networks.

Intrusion is an attempt to compromise the integrity, availability, and confidentiality of the network security system or to bypass the security mechanisms of a computer system or network. Infact, it is an illegal access to the network system.

the plaintext is divided into two halves. The DES form of standard takes input as 64-bit plain text and 56-bit key to produce 64-bit ciphertext.

There are differences between the DES and the AES standards. The basic difference between the two is that the block in DES is divided into two halves before further processing, whereas in AES, the entire block is processed to obtain ciphertext. The DES algorithm works on the Feistel Cipher principle, and the AES algorithm works on substitution and permutation principle.

Intrusions can happen in different ways like a system being infected with malwares like worms and spywares, attackers gaining unauthorized access to systems from the Internet, and authorized users of systems who misuse their privileges or attempt to gain additional privileges for which they are not authorized. Although many intrusions are malicious in nature, many others are not; for example, a person might mistype the address of a computer and accidentally attempt to connect to a different system without authorization.

Intrusion detection is the process of monitoring the events occurring in a computer system or network and analyzing them for signs of possible intrusions (incidents).

Intrusion detection system (IDS) is software that automates the intrusion detection process. The primary responsibility of IDS is to detect unwanted and malicious activities.

Intrusion prevention system (IPS) is software that has all the capabilities of an intrusion detection system and can also attempt to stop possible incidents.

11.5.1 The Need for Intrusion and Prevention Systems

Intrusion and detection systems are very much needed in network security systems for the following reasons:

- They help to identify problems with security policies.
- They help to document existing threats.
- They identify possible incidents, logging information about them, attempting to stop them, and reporting them to security administrators.
- They help deter individuals from violating security policies.
- They help to identify possible incidents, logging information about them, attempting to stop them, and reporting them to security administrators.
- They are known to serve as one of the cost-effective ways to block malicious traffic.
- They help to detect and contain worm and virus threats, to serve as a network monitoring point, to assist in compliance requirements, and to act as a network sanitizing agent.

11.5.2 Implementation of IDS Technique

Implementation of this IDS technique involves two major parts: "set-up inside the network" and "set-up at the central detection point." The following should be the steps under the "set-up inside the network" part:

- Group all the nodes in the network into various IDS circles (i.e., each IDS circle consists of one central IDS or IDSC and all the peripheral IDSs or IDSPs linked to it).

- Install a detector (i.e., software on a computer) at each IDSP for determining the packet arrival rate, etc. at that node.
- Install a transmitter at each IDSP for sending the packet arrival rate data, etc. to the central detection point by multi-resolution techniques.

The following should be the steps under the "set-up at the central detection point" part:

- Install OPNET 14.0 at the central detection point (i.e., software on a computer).
- Replicate the actual network in OPNET 14.0 at the central detection point with reference to Fig. 11.5 or the actual implementation scheme for the IDS.
- Install a receiver to receive the packet arrival rate data, etc. from the network (i.e., at the end of the multi resolution techniques).
- Furnish the replica of the actual network developed in OPNET with the packet arrival rate data, etc. of all the IDSP nodes.
- Carry out all the steps discussed under the simulation studies above.

Figure 11.5 shows the overview of the implementation scheme for the automated IDS technique. Only two IDSC nodes can be efficiently implemented for this sample network. Each IDSC node is linked to four IDSP nodes for detection. Please note that LAN 1 does not form part of the IDS nodes because it is not linked directly to any of the chosen central IDS nodes. All security data from the IDSC and IDSP nodes will be directly transmitted to the central detection point for analysis. This is done using the multi-resolution techniques.

11.6 Cybersecurity

The technological process by which computers, networks, programs, and data can be protected from unauthorized access or attacks that are aimed to collect information or cause harm maliciously is known as cybersecurity or information technology security.

Cybersecurity is the practice of defending electronic systems, data, mobile devices, computers, networks, and servers from malicious attacks.

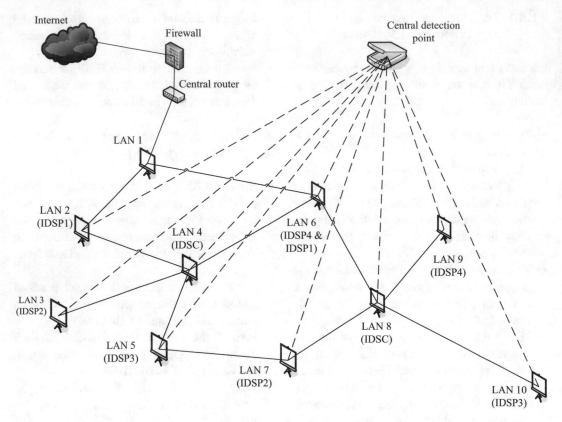

Fig. 11.5 Overview of implementation scheme for the automated IDS

Network security includes activities to protect the usability, reliability, integrity, and safety of the network. Cybersecurity can be classified into several domain areas such as:

- Cyber threat (malware, virus, intrusion, DoS) detection and protection
- Internet of Things (IoT)
- Cloud computing
- Communication security (wireless and wired) in social networking and virtual environments
- Artificial intelligence in cybersecurity
- Cybersecurity in power grid and smart technologies
- Blockchain
- Cybersecurity education

11.6.1 Mitigation Against Cybersecurity

The solution to the issues concerning cybersecurity can be solved in different ways:

- *Virus and malware:* The solution in this case requires different experimentations with and staying abreast of latest viruses and malware.
  - Using an isolated network for deliberately inducing infections
  - Understanding attack vectors
  - Establishing coordination with other cybersecurity labs
  - Monitoring evolving threats
  - Understanding zero-day attacks
- *Virtual environments:* Cybersecurity attacks in virtual environments can be mitigated using hypervisors such as:

– Type 1: Bare metal
– Type 2: Hosted

The configuration can be by virtual machines (VM). The operating systems can be Windows and Linux virtual machines. It uses software-defined networking.

- *Wireless security:* The cybersecurity solution in the wireless can be Wi-Fi focused. The spectral protection can be accomplished by:
  – Resilience against interference and jamming
  – Encryption and authentication schemes
  – Identity management
- *Cloud security:* The mitigation of the cloud security issues can be done by using:
  – Cloud-based firewalls
  – Virtual private cloud
  – Multi-cloud security architectures
  – Simultaneous use of heterogeneous clouds
  – Cloud backup and collaboration
  – Backups
  – Sharing
  – Encrypted email
  – Secure links instead of attachments
- *IP networking exploits:* The possibility of mitigating against IP networking exploits can be done by the process of:
  – Denial of service (DoS) attacks
  – DDoS
  – Application overload
  – Selective or malformed queries
  – TCP Syn
  – Ping flood
  – Smurf
  – IP address spoofing
  – DNS spoofing
  – Cache corruption
- *Cryptography and Secure Socket Layer (SSL):* In this case, the solutions can be carried out using processes as follows:

  – SSL encryption as a boon
    (a) For the bad guys!
    (b) Allowing concealment of traffic
    (c) Traditional firewalls and proxies ineffective

– Focus areas
  (a) Encryption
  (b) Authentication
  (c) SSL certificates
  (d) Certificate chaining
  (e) Public key cryptography
– SSL inspector

Summary

1. Network security is very much needed by users and more so by businesses.
2. Malwares are always very harmful, destructive, intrusive, and hostile, with the goal of trying to destroy, disable, invade, or damage the computer network.
3. The packet filter is used to filter incoming or outgoing interfaces such as ingress filtering of spoofed Internet Protocol (IP) addresses and egress filtering.
4. Firewalls offer a convenient point where Internet network security can be monitored and alarms generated.
5. Encryption is the mechanism used to ensure that the content of the information from point A (source) to point B (destination) still retains its original content and integrity.
6. The Data Encryption Standard (DES) is the most widely used standard for symmetric encryption.
7. The AES is comparatively faster than DES.
8. A digital certificate is a unique identifier assigned to a user by a certification authority to verify the identity of the user.
9. All security data from the IDSC and IDSP nodes are directly transmitted to the central detection point for analysis.
10. Network security includes activities to protect the usability, reliability, integrity, and safety of the network.

Review Questions

11.1. The different types of malware include spyware, computer viruses, Trojan horses, and worms.

(a) True   (b) False

11.2 Types of malware include:

(a) Botnets   (b) Adware   (c) a and b

11.3 Installing and using anti-malware software is not one of the best ways to take care of malwares.

(a) True   (b) False

11.4 The packet filter is not one of the firewall mechanisms that are used to protect the network of an organization.

(a) True   (b) False

11.5 TCP means:

(a) Transmission Control Protocol (b) Transmission Control Produce

11.6 UDP means:

(a) User Dedicated Protocol   (b) User Datagram Protocol

11.7 In the encryption implementation, the key must not be kept secret so that the security of the network is not compromised.

(a) True   (b) False

11.8 Intrusion and detection systems are not needed in the network security system.

(a) True   (b) False

11.9 The two main functions of the digital signature are:

(a) To identify the signer and the approbation of the document to be transmitted

(b) To identify the malware and the document to be transmitted

11.10 The cybersecurity solution in the wireless can be Wi-Fi focused.

(a) True   (b) False

Answer: 11.1 (a), 11.2 (c), 11.3 (b), 11.4 (b), 11.5 (a), 11.6 (b), 11.7 (b), 11.8 (b) 11.9 (a), 11.10 (a)

Problems

11.1 List at least seven common types of malware.

11.2 Describe briefly the functions of the malwares listed in Problem 11.1.

11.3 List at least ten different anti-malware removal software tools.

11.4 Briefly describe each of the ten different anti-malware removal software tools listed in Problem 11.3.

11.5 (a) Define a firewall security system. (b) What are some of the benefits of having a firewall for network security?

11.6 What are the limitations of a firewall system?

11.7 (a) What is encryption? (b) How can the encryption and decryption of information be implemented in a network system?

11.8 If we assign each of the 26 letters of the alphabet to a unique integer from 1 to 26, using the encryption rule of additive cipher mod 25, encrypt the word "happy."

11.9 Briefly describe the three basic types of encryption.

11.10 What is symmetric/private key encryption?

11.11 What is asymmetric/public key encryption?

11.12 What is the function of a digital certificate?

11.13 What are the two main functions of a digital signature?

11.14 (a) What is intrusion in a network system? (b) How can intrusion happen in a network system?

11.15 Why are intrusion and detection systems needed in network security systems?

11.16 What are the various domains cybersecurity can be classified?

Emerging Technologies

12

U.S. computer networks and databases are under daily cyber attack by nation states, international crime organizations, subnational groups, and individual hackers.

John O. Brennan

Abstract

The advancements in the Internet technologies and connected smart objects have contributed in making Internet of Things (IoTs) a widely pervasive computing. There has been a complete evolution of mobile networks from 1G network to now emerging 5G network. The 5G network is an end-to-end ecosystem to enable a fully mobile and connected society. In this chapter, we cover many of the emerging technologies such as blockchain, IoTs, big data, smart city technologies, and cybersecurity issues.

Keywords

Internet of Things · Blockchain · 5G technology · Cloud computing · Edge computing · Fog computing

12.1 Internet of Things

As new technologies continue to emerge as shown in Fig. 12.1, we find that it has become possible to use smart technologies such as your cell phones to control devices in your home, office, and car, monitor what people are doing at any moment in time, or even monitor service events. These devices are interconnected with the help of the Internet to the cell phones as an example to perform specific functions as shown in Fig. 12.2.

Internet of Things (IoT) is the interconnection of smart objects or things. When interconnected to people, it becomes Internet of People (IoP), and when interconnected to service, it becomes Internet of Service (IoS).

The IoT could be defined as a network of -Internet enabled devices, machines, objects, and people which work together autonomously by gathering surrounding data, performing analysis on it, and providing valuable information to the IoT stakeholders along with accomplishing a specific task or function.

IoT could also be defined as a network of sensors and actuators embedded into physical objects which has the ability to connect to the Internet and gather and exchange data with other objects/things, people, machines, and systems in the network.

Gartner Hype Cycle for Emerging Technologies, 2017

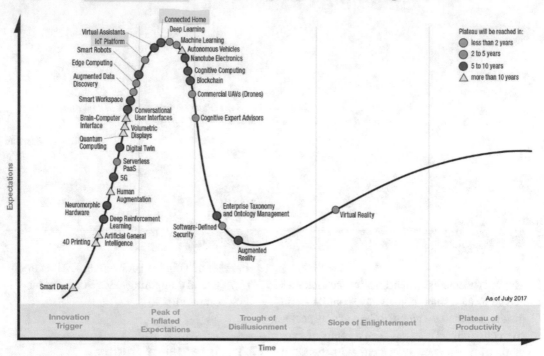

Fig. 12.1 The Gartner hype cycle for emerging technologies, 2017. (Source: https://www.gartner.com/smarterwith-gartner/top-trends-in-the-gartner-hype-cycle-for-emerging-technologies-2017/)

Fig. 12.2 The Internet of Things (IoT). (Source: http://www.sitel.com/blog/revolutionizing-customer-service-with-the-internet-of-things/)

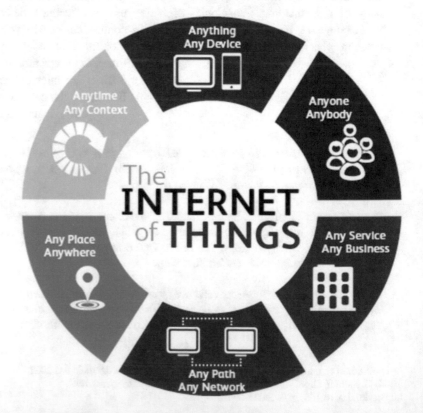

These "things" or "devices" are capable of autonomously gathering data and information from their surroundings, which can be used in accomplishing a specific task. Recently, IoT has gained rapid momentum and is being explored in every field, such as commerce, government, academia, and industrial applications. The advancements in the Internet technologies and connected smart objects have contributed in making IoT a widely pervasive concept. IoT applications and services have tremendously changed our lives and will continue to do so in the future. However, there are many challenges associated with IoT, such as:

- Limited resources (e.g., power, storage, etc.)
- Scalability
- Heterogeneity
- Security
- Privacy

Many researchers and industry players are addressing these challenges with significant research and innovative initiatives for continued success of IoT. Cloud computing is one of the most important enabling infrastructures for IoT.

12.1.1   History of Internet of Things

- 1982: An Internet-connected coke machine was built at Carnegie Mellon University for inventory management and query the coldness of coke.
- Early 1990s: Research and publications in ubiquitous computing by Mark Weiser's concept of connected home appliances by Reza Raji.
- Late 1990s: Device-to-device communication first introduced by Bill Joy.
- 1999: Kevin Ashton, founder of Auto-ID Center at MIT, first coined the term Internet of Things and identified Radio-Frequency Identification (RFID) technology as the foundation of IoT.
- 2008–2009: The number of connected devices surpassed the human population.

- 2011: IPV6 launched, one of the enabling technologies for identifying billions of smart things, and Gartner included Internet of Things in the hype cycle of emerging technologies (reached the peak of cycle around 2014)
- 2014–Present: Big companies started to take IoT initiatives: Cisco, IBM, Amazon, etc.
- 2014–Present: Commercial hardware platforms for IoT (e.g., Arduino) and IoT software platforms (e.g., IBM Watson), IoT (using blockchains), Amazon AWS IoT, Azure IoT Suite, etc.

12.1.2   Evolution of Internet of Things

The evolution of IoT started with the following ideas and technologies:

- Concept of IoT gained momentum with advancements in *wireless technologies and Radio-Frequency Identification* (RFID) technology.
- Public Internet and World Wide Web-enabled machines to talk to each other.
- Number of connected devices increasing at an *exponential* rate.
- Launch of IPV6—unique identification of things in virtual space.
- IoT architectures: three basic layers—*object/ perception layer*, *middleware layer(s)*, and *application layer* as shown in Fig. 12.3.
- Recently, smart devices developed and deployed in the industry such as NEST (smart thermostat, light bulbs) and connected home appliances (refrigerator, cameras, etc.).
- With cloud computing capabilities, IoT platforms were introduced by major cloud services providers (e.g., Amazon, IBM, Google, etc.).
- *Internet of Everything (IoE)* by CISCO: *networked connection of people, process, data, and things.*
- *Network of things (NoT)* by NIST: *model is based on four fundamentals at the heart of IoT—sensing, computing, communication, and actuation.*

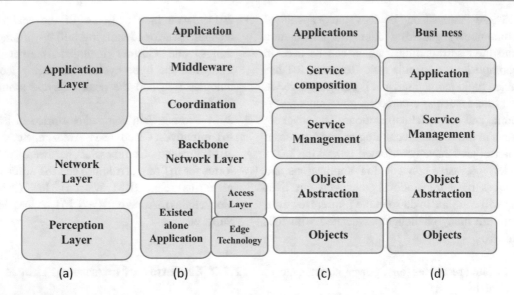

Fig. 12.3 The IoT architecture. (**a**) Three-layer. (**b**) Middle-ware based. (**c**) SOA based (**d**) Five-layer

- It has been predicted by the International Data Corporation that the global market for IoT solutions will grow to about $1.7 trillion by 2020.

IoT is evolving with "anything" and "everything" being connected to the Internet as well as becoming smarter with support of technologies like:

- Cloud computing
- Artificial intelligence (AI)
- Big data analytics
- Machine learning

According to Gartner hype cycle for emerging technologies as of July 2017 as shown in Fig. 12.1, IoT platforms will move toward the peak of the cycle and are expected to reach the plateau within the next 2–5 years. Influenced by these predictions and advancements in the IoT arena, numerous IoT applications and services are being developed and deployed in different sectors, such as:

- Industry
- Government
- Education
- Transportation

- Infrastructure
- Healthcare
- Military

Some of the common applications of IoT are:

- Connected smart homes
- Connected appliances (e.g., thermostats)
- Smart cities
- Connected cars
- Smart grids
- Smart medical devices and applications
- Wearable devices, etc.

12.1.3 Examples of Various Internet of Things

IoT is applied in various areas of life such as shown in Fig. 12.4:

- *Transportation and Logistics*
  Sensors enabled traffic management, mobile ticketing, smart parking, and inventory and supply chain management.
- *Healthcare*
  Smart medical devices, remote patient monitoring, and ambient assisted living for elderly and chronic disease patients.

Fig. 12.4 IoT application domains. (Source: Bhatt, Smriti, Farhan Patwa, and Ravi Sandhu. "An Access Control Framework for Cloud-Enabled Wearable Internet of Things." In 3rd *International Conference on Collaboration and Internet Computing (CIC)*, pp. 530–538. IEEE, 2017)

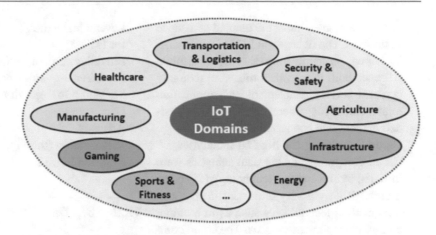

- *Infrastructure*
  Smart cities, smart grids, and utility maintenance.
- *Security and Safety*
  Wearable security and safety devices.
- *Sports and Fitness*
  Fitness trackers, smart clothing and accessories, and smartwatches.
- *Agriculture*
  Sensors enabled for watering plants and ideal crop yielding.
- *Industrial and Manufacturing*

  For energy preservation, IoT sensors are utilized to monitor energy usage and data.

Other examples of IoT that have already been realized in different subfields are:

- Wearable Internet of Things (WIoT)
- Industry Internet of Things (IIoT)
- Internet of Vehicles (IoV)
- Vehicular Internet of Things (VIoT)
- Medical Internet of Things (MIoT)

12.1.4 IoT Enabling Technologies and Services

In IoT infrastructures and services, there are various enabling technologies and cloud-enabled platforms that help to make things happen. These are:

- IoT Enabling Technologies

- IoT devices cannot function without the key technologies that make them work. That means that IoT is enabled by several technologies including cloud computing, wireless sensor networks, embedded systems, communication protocols, big data analytics, mobile Internet, semantic search engines, security protocols and architectures, and web services.
- Sensors and Actuators
- A sensor is the same as a transducer. Any physical device that converts one form of energy into another is called a transducer. In the case of actuators, they operate in the reverse direction of a sensor. Actuators take electrical inputs and turn them into physical actions.
- Ubiquitous Internet, Wireless Technology (e.g., Zigbee, Lightweight Bluetooth, 6LoWPAN), and RFID Technology
- In the ubiquitous Internet, Wireless Application Protocol (WAP) is used which is a specification for a set of communication protocols to standardize the way that wireless devices can be used for Internet access.
- M2M Communication and Networking Protocols (e.g., MQTT, CoAP, AMQP, etc.)
- M2M is a machine-to-machine communication and networking protocol that is used to describe any technology that enables networked devices to exchange information and

perform actions without the manual assistance of humans. The lightweight M2M is the industry open protocol from the Open Mobile Alliance built to provide a means to remotely perform service enablement and application management for "Internet of Things" embedded devices and connected appliances.

- Cloud Computing and Big Data Analytics
- Cloud computing and big data analytics were initiated by information technology (IT) community. Both are very important in many of the enabling technologies such as in building energy management analysis. The cloud computing has its own advantages because of its virtualized resources, parallel processing, security, and data service integration with scalable data storage.
- Smartphones and Mobile Computing
- Smartphones are cell phones that allow the user to do more than calling on the phone or sending text messages. Smartphones can be used to browse the Internet and run software programs like a computer. They use touch screens to allow users to interact with them. Mobile computing is a "human-computer interaction" that transports data, voice, and video over a network via a mobile device. These mobile devices can be connected to a local area network (LAN), and they can also

- Leveraging of the virtually unlimited resources for IoT
- Providing secure bidirectional communication: *device-to-cloud* and *cloud-to-device*
- Managing IoT security
- Data storage
- Analysis
- Visualization (through application domains as shown in Fig. 12.4)

12.2   Big Data

The term big data refers to the collection of an immense amount of data that is collected every second of every minute of every hour of every day. The collection of such data is always a continuous process. The amount of such data is hard to store and hard to analyze. In addition to the analysis of the data, there is the need to utilize the data in order to improve systems and operational functions. We will explore what big data is and how it came to be as well as how big data can be applied to different concepts in the next section.

12.2.1 What Is Big Data?

Big data is the collection of data from a multitude of sources from anything done digitally.

take advantage of Wi-Fi or wireless technology by connecting through a wireless local area network (WLAN).

- Others: Cloudlets, Fog Computing, Edge Computing, and Dew Computing

- Cloudlets, fog computing, edge computing, and dew computing are computing models proposed to provide some features that cloud computing cannot provide. They share one common feature: they all perform computing tasks at devices that are closer to users. They can be called post-cloud computing models.

In the cloud-enabled IoT platforms, the following benefits can be expected:

The collection of data has grown exponentially over the years. In 2003, 5 exabyte of data were created by people in a year. It is hard to imagine that that size of data can now be created within 2 days. In 2012, the digital world expanded to 2.72 zettabytes of data. The sheer amount of data that are created every day is mind-boggling. The human face of big data is a global project that centers on the creation of real-time data collection and analysis of large data. This media project deduces many statistics such as Facebook having 955 million active monthly accounts with 70 languages being used. Approximately 140 billion photos are uploaded on a monthly basis by just Facebook. When it comes to sites like YouTube, you will find that for every 60 s, 48 h of video are uploaded and 4 billion views take place every day. Data can also be extracted from sending an email to a friend and other digital sources that we have at our disposal today and that will be made available in the future. As an example, a sensor in the Fitbit Blaze has the ability to keep track of the numbers of steps taken in a day, the sleeping pattern of the individual and what hours that person gets the most productive sleep, how many times the person wakes during the night, the amount of water the person drinks, and many other features that can be of use to the user.

Internet of Things is a term and concept that was introduced in 1999, and the basic concept is that everything can be interconnected, communicate, and function with limited interference from people.

Self-driving cars, printers that can order their own toner, and vehicles that can call ahead to let others know you will be late are all ideas of IoT, and some have been implemented, while others are still work in progress and will be combatted in the future. All of these innovations are made possible due to big data. The collection of data has grown exponentially over the last 10 years and will continue to grow for the foreseeable future. While all of this is good, there are several drawbacks to big data. Some of these drawbacks are as shown in Fig. 12.5:

- Variety
- Velocity
- Volume

There are a variety of different fields that utilize big data to assist in the careful analysis to gain insight and depths to help solve real-life problems. Variety is what makes big data even bigger. Big data is collected from a variety of different sources and it comes in three different types. Those types are known as structured, semi-structured, and unstructured.

Velocity is something that is not only required for big data but for processes in general. The goal is to process the data as it is streamed in order to maximize the value for a given organization. The volume aspect has exceeded previous amounts of bytes so new levels have had to be created. The

amount of data created has outgrown traditional storage and analysis techniques.

12.2.2 The Evolution of Big Data

Data have always been collected on people and what they are doing. However with big data, everything down to the most miniscule component can be analyzed and used to maximize the outcome of a person's experience or even a company's profits. The concept of big data has been developing over the last 100 years.

- In 1997, Michael Lesk publishes a paper titled "How Much Information Is There in the World?" where he theorizes about the existence of 12,000 petabytes. In addition to this, he also points out that at this early stage of the web's development, it was increasing 10-fold every year. This was also the year that Google made its debut.
- In 1999, the term big data comes into play, thanks to an article published in a magazine titled "Visually Exploring Gigabyte Datasets in Real Time" by the Association of Computing Machinery. In the article, the problem of storage was again brought up. This was also the first time that the term "Internet of Things" was used. RFID pioneer Kevin Ashton used the term for a presentation that was given to Procter and Gamble.
- In 2000, Peter Lyman and Hal Varian made the first attempt to determine exactly how much data was in the world and the rate that it grew. They were able to conclude that "The world's total yearly production of print, film, optical and magnetic content would require roughly 1.5 billion gigabytes of storage. This is the equivalent of 250 megabytes per person for each man, woman and child on Earth."
- In 2001, Doug Laney defined the three characteristics of big data in his paper "3D Data

Fig. 12.5 The three V's
of big data. (Sagiroglu
and Sinanc 2013)

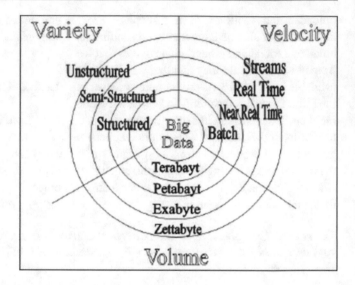

Management: Controlling Data Volume,
Velocity and Variety." This was also the first
time that the term "software as a service" was
used. This term was important because it was
a fundamental concept in today's cloud-based
applications.

- In 2005, Web 2.0 was announced which meant
  that instead of content provided by services
  providers, user-generated content was on the
  rise. Hadoop was also created about the same
  time.
- In 2007 the concept of big data was brought to
  the masses through an article by Marr.
- In 2008, 9.57 zettabytes of information were
  processed on servers around the world.
- In 2010, Eric Schmidt of Google speaks on

ing device to access digital data than a
home or office computer. Also 88% of businesses made big data analytics their first
priority.

12.3 Smart Cities

The term "smart cities" may be defined differently by many people depending on what they
perceive to be what makes them comfortable
based on technology enhancements. The term has
exploded in the last decade as new technologies
and devices have made it possible to integrate
and facilitate urban living in ways we never
imagined.

*Smart cities are cities that have efficient data-driven infrastructures that use technologies to
enhance the functions of the cities while at the same time increasing the quality of live.*

how the amount of data being created in
2 days currently is what was created from the
beginning of the human race until 2003.
- In 2011, McKinsey's report stated that by
  2018 there would be a shortage of professional
  data scientists and that issues concerning privacy, security, and intellectual property had to
  be fixed before the full potential of big data
  could be achieved.
- In 2014, it was reported that more people
  were using their phone or any other travel-

In building a bridge to a smart city, you have
to make sure that everything including all services, all data, and all devices can be seamlessly
connected. Every smart city starts with its network. It must be noted that fast-growing cities
across the world are faced with proactively solving future challenges. As access to broadband
network and high-speed gigabit fiber rises, there
are increasing opportunities to develop the next-
generation applications for smart urban planning. Not only can these advanced technology

solutions predict future obstacles, they enable new ways for citizens to engage with government.

Designing a smart city is a monumental task. To have a real smart city, the following elements must be present as part of the city and as depicted in Fig. 12.6.

- Improved city efficiency
- Smarter digital infrastructure
- Healthcare services
- Citizen engagement and digital citizen services
- Smart energy
- Smart transport
- Public safety and resiliency
- Leadership and vision
- eGovernance
- Smart, big, and visualized data
- Citizen satisfaction
- Improving quality of life of the citizenry
- Economic vitality
- Sustainability
- Relying on data and leveraging technology, especially information communication technologies throughout the city.

A typical example of the key elements that make up a smart and connected city could also be seen pictorially in Fig. 12.7. It starts with identifying the enabling technologies and finding which districts the technologies will be deployed and the possible outcomes. These outcomes could include improvement in safety, enhancement of mobility, enhancement of upward climbing of the ladders of opportunities, and how all these address the effects on the change on climate.

A second example, as depicted by James Staubes of Symmetry Electronics, is shown in Fig. 12.8 demonstrating the future farms, small and smart. In the smart farm, you have the survey drones, fleet of agribots, farming data, texting cows, and smart tractors.

Survey drones: It has aerial drones that survey the fields, mapping weeds, yield, and soil variations. This enables precise application of inputs, mapping spread of pernicious weed blackgrass resulting to substantial increase in wheat yields.

Fleet of agribots: These are a herd of specialized agribots that tend to crops, weeding, fertilizing, and harvesting. The robots are capable of

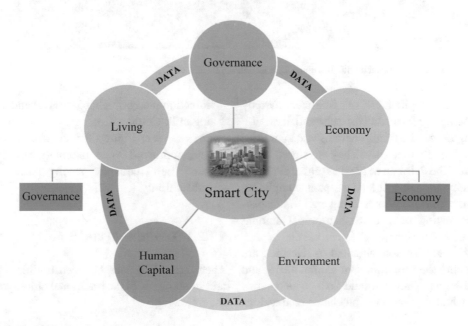

Fig. 12.6 Elements of a smart city

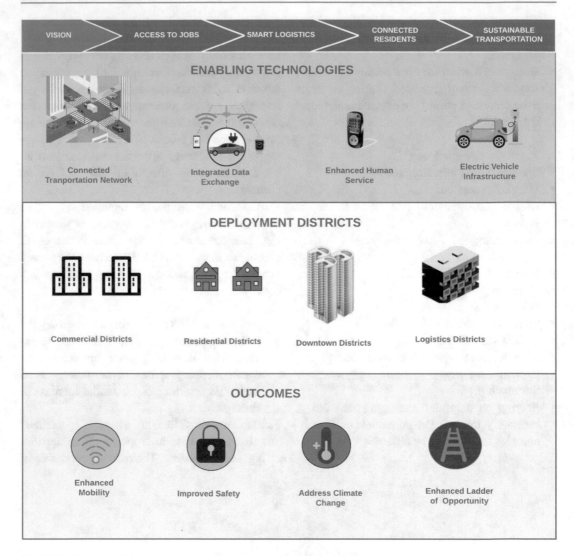

Fig. 12.7 Example of the elements that make a smart and connected city

microdot application of fertilizer thereby reducing the fertilizer cost substantially.

Farming data: The farm generates vast quantities of rich and varied data. They are stored in the cloud. The data can be used as digital evidence thereby reducing the time spent completing grant applications or carrying out farm inspections saving on average substantial amount per farm per year.

Texting cows: Sensors attached to livestock are allowing the monitoring of animal health and well-being. They can send texts to alert farmers when a cow goes into labor or develops

infection increasing herd survival and increasing milk yields.

Smart tractors: Using GPS-controlled steering and optimized route planning, soil erosion can be reduced saving fuel costs substantially.

12.3.1 Traditional and Smart Cities

There are differences between traditional cities and smart cities. In the traditional cities, you have the following:

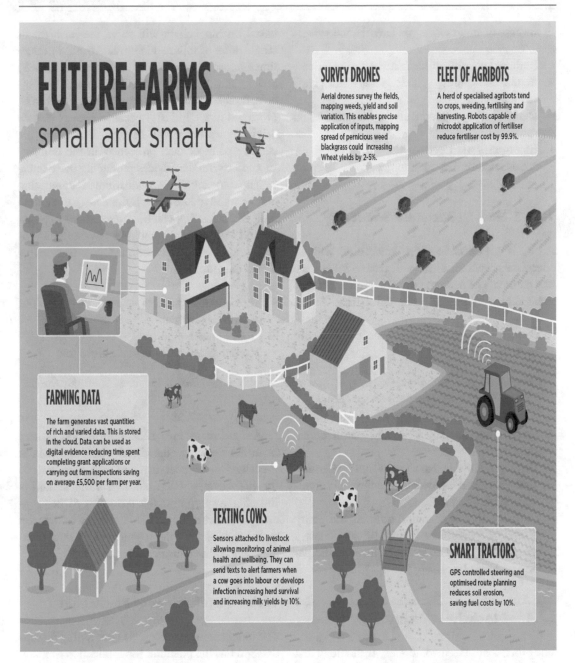

Fig. 12.8 Future smart farms. (Source: Symmetry Electronics, https://www.semiconductorstore.com/pages/asp/DownloadDirect.asp?sid=1548064503461)

- High utility bills and inefficient use of resources
- Unsolved theft and crime
- Limited records/information
- Difficult communication challenges, especially in emergencies
- "Band-Aid" repairs to infrastructure

In the case of smart cities, you have the following:

- Reduced cost of utilities due to optimization
- Increased safety and tools for law enforcement
- Data storage/increased communication resources
- Emergency alert systems
- Infrastructure that communicates

12.3.2 How to Become a Smart City

A city is not classified as a "smart city" without adequate preparation that will enable it have the key elements of a smart city. The preparation has to be abided by the following guidelines:

- Identifying key citizen centric issues to resolve.

Blockchain is a cryptographically secured system with a distributed ledger that allows for the secure transfer of data between parties.

- Developing a plan with key partners and stakeholders.
- Finding solutions that address multiple issues.
- Collecting and using data to improve technologies benefiting citizens.
- Collaboration is essential: utilizing university research, creating public-private-industry partnerships, and cultivating citizen participation and buy-in.

12.4 Blockchain Technology

The blockchain is a revolutionary technology that started with cryptocurrency bitcoin in 2008. Today, blockchain is impacting numerous fields of endeavors such as medicine, manufacturing, and supply chain to name a few with several use cases. Internet of Things (IoT) must be fully reliable and trusted in the industry to realize its full benefits. The current server-client model for the IoT is not fit for the healthcare industry due to high risk associated with a single point of failure. A distributed or peer-to-peer model for the IoT

based on blockchain will solve the problem of single point of failure and provide trusted protection against distributed denial of service (DDoS) attacks that use vulnerable IoT devices as botnets. It makes sense to discuss how the IoT can be protected in the industry because numerous data originate from the IoT devices. Therefore, the starting point of data protection in the industry should be with the devices producing these data.

The blockchain is a distributed computing platform and secure by design and has a Byzantine fault tolerance that can be programmed to transfer and store or share any item of value in a process known as a transaction. Other features of blockchain are:

- The consensus process achieved through proof of work algorithm
- Proof of stake algorithm
- The delegated proof of stake algorithm

As shown in Fig. 12.9, information is sent from the device to the cloud where the data is processed using analytics and then sent back to the IoT devices. Each block in a blockchain is linked using cryptographic technology. The properties of blockchain can be leveraged to develop a secure model for the IoT as well as data management in the industry.

12.4.1 Applications of Blockchain

The blockchain technology is new, highly disruptive and has the potential for application in many fields of study. The blockchain, therefore, is the focus of many researchers today as an emerging new technology. In the field of medicine as well as manufacturing, the blockchain promises to be a game changer due to its potentials and benefits. Therefore, it is paramount that a blockchain customized for medical applications meets certain needs in medicine. To efficiently handle electronic medical records (EMRs), a distributed record management system using blockchain

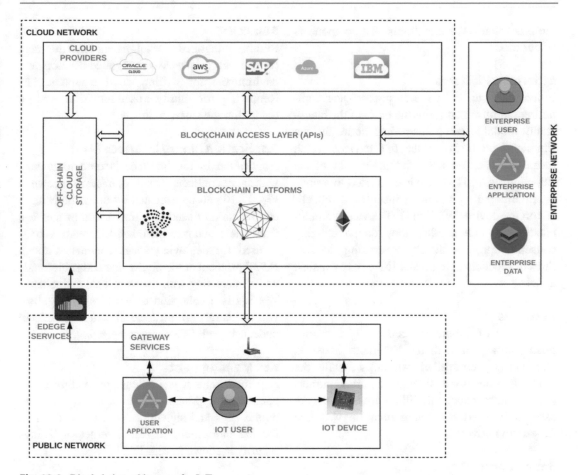

Fig. 12.9 Blockchain architecture for IoT

concepts and features can be developed. A comprehensive discussion on the potential of blockchain for medical research, healthcare management, and drug counterfeiting has been implemented.

Blockchain has the potentials and economic benefits of solving the problem of scale in the healthcare industry. The sensitivity that medical data are expected to have can be guaranteed by the privacy and transparency that can be offered by using blockchain technology. The application of efficient key management scheme can be implemented as a mechanism for enhancing the privacy of data in the medical blockchain. The problem in medical data management can be solved using blockchain for the provision of secure and scalable data exchange. Traditional blockchain architectures have been reworked and improved and can be used to support the func-

tionalities of medical blockchain like parallel computing, data management, identity management, and data sharing in a trusted, verifiable, and anonymous way.

12.4.2 Customized Blockchain Applications

There is still an important need for a customized blockchain for deployment in the IoT-enabled application architectural IoT environment as shown in Fig. 12.9.

Software and Hardware Tools

Internet of Things devices applicable to the field of medicine can be securely managed using blockchain. In this section, we present how the medical Internet of Things can be managed using the blockchain. We first discuss the components of our model.

Ethereum Platform

Ethereum is an open-source peer-to-peer computing platform that forms the basis of the blockchain model. The participating nodes in the Ethereum platform are the IoT devices which mean that each device maintains a copy of the Ethereum. An IoT device is converted to a node on the Ethereum platform by installing a geth client on the device. When the IoT devices communicate with one another, they do so utilizing consensus process thereby leveraging the features of blockchain to protect IoT devices against attacks.

Accounts

Two types of Ethereum accounts can accomplish interaction with the Ethereum network. One can be controlled externally, while the other is controlled through smart contracts (smart contract account). Ethereum accounts are essentially used for communication with the Ethereum network.

Smart Contracts

A model using blockchain can be composed of multiple self-executing and self-enforcing contractual logics called smart contracts. Smart contracts can be written for each of the IoT devices on the Ethereum platform for controlling the device and saving policy values of devices. Signature and public key in a smart contract ensure that a valid is successfully authenticated.

Remix IDE

Remix Integrated Development Environment (IDE) is a browser-based integrated development environment for writing smart contracts. The Remix IDE for solidity is used for writing and for compiling the smart contracts.

Application Binary Interface (ABI)

Application binary interface is used to interact with smart contracts. Smart contract ABI shown in Fig. 12.10 can be executed by either invoking a call function or sending a transaction or message from one account to the other. Ethereum/web3.js is used for deploying smart contracts through HTTP protocol at the object side and subject side while interacting with the associated geth clients. Web3.js is a collection of libraries which allow you to interact with a local or remote Ethereum node, using a HTTP or IPC connection.

Key Management

Key management is implemented within smart contracts for proper validation of inputs using public keys and signatures. The block diagram for the blockchain for the Internet of Things security is shown in Fig. 12.11.

12.5   Cloud Computing

The idea of cloud computing started about 20 years ago. The technologies and issues involved in cloud computing include fog computing, edge computing, the Internet of Things (IoT), quality of service, forensics in cloud, and standards.

When computing services such as networking, databases, storage, intelligence, servers, analytics, software, and many more are delivered over the internet ("the cloud") in such a way as to facilitate better savings, flexible resources, and much faster innovation, we refer to such computing services as cloud computing.

The access control contract, judge contracts, and register contracts are necessary for distributed and reliable access control for the Internet of Things in the field of medicine where sensitive medical data need to be secured.

This kind of computing is very useful in the sense that you pay only for the cloud services that you use. It helps in running your business or research infrastructure efficiently and helps in operating costs savings.

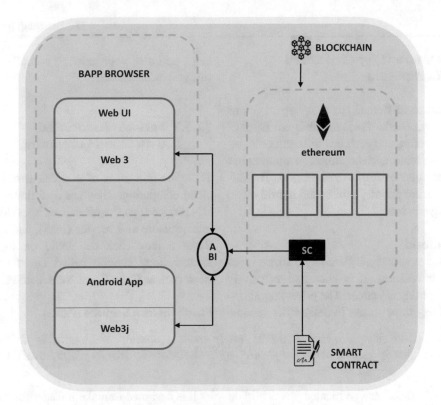

Fig. 12.10 Application binary interface (ABI)

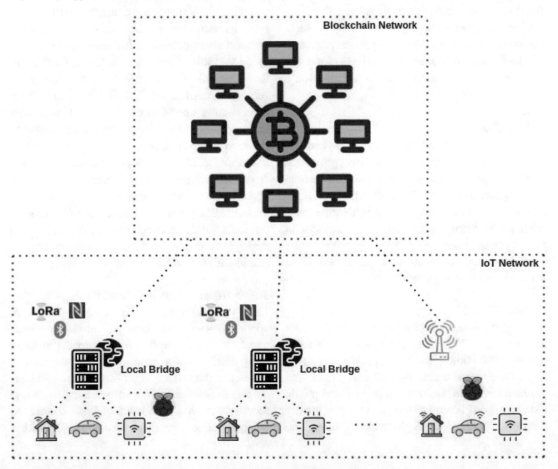

Fig. 12.11 Diagram for the blockchain IoT security. (Source: https://www.mdpi.com/1424-8220/22/4/1411/htm)

12.5.1 Different Types of Cloud Computing

In real world, different business organizations have different needs. There are different types of cloud computing and each has different configurations. Therefore, each organization must choose which one is better for its organization. The different types are private, public, and hybrid cloud computing services.

Private Cloud

If the cloud is specifically used by a single organization or business, it is classified as private cloud computing resources. The private organizations can pay third parties to deliver the service, or the private cloud can be located physically on the organization's on-site data center. A private cloud is one in which the services and infrastructure are maintained on a private network. Some companies also pay third-party service providers to host their private cloud. In a private cloud, the services and infrastructure are maintained on a private network.

Public Cloud

In the case of public clouds, they are owned and operated by third-party cloud service providers. They deliver their computing resources like storage and servers over the Internet. All of software, hardware, and other supporting infrastructure is owned and managed by the cloud provider in public cloud computing. Web browser is the available possible means you can access these services and manage your account.

Hybrid Cloud

In hybrid clouds, the private and public clouds are combined. They are bound together by technology that allows data and applications to be shared between them. By allowing data and applications to move between private and public clouds, a hybrid cloud gives your business greater flexibility and more deployment options and helps optimize your existing compliance, security, and infrastructure.

12.5.2 Services Associated with Cloud Computing

There are four areas of services associated with cloud computing. They are as follows: platform as a service (PaaS), software as a service (SaaS), infrastructure as a service (IaaS), and serverless. Because they each can build on top of one another, they can be called cloud computing stack such as CloudStack and OpenStack.

Platform as a Service (PaaS)

The cloud computing services that supply an on-demand environment for delivering, developing, managing, and testing software applications are known as platform as a service (PaaS).

It is designed to make it easier for developers to quickly create mobile apps or web, without worrying about managing or setting up the underlying infrastructure of databases, network, servers, and storage, needed for development. PaaS uses OpenShift, Cloudify, and Cloud Foundry in its operations. The OpenShift open-source software is developed by Red Hat. It leverages on several other open-source software. It also adds Development Operations (DevOps) tools to improve deployed applications' development and maintenance. In the case of Cloudify, it is a topology and orchestration specification for cloud applications (TOSCA) and services. It automates the entire lifecycles. The Cloud Foundry is a self-service application execution engine, automated deployment engine, and life cycle manager integrated with various development tools.

Software as a Service (SaaS)

This type of service is for delivering software applications over the Internet, on demand used mostly on a subscription basis. Cloud providers using SaaS host and manage the software application and the underlying infrastructure and handle any maintenance, like software upgrades and security patching. The users of SaaS connect to the application over the Internet, usually with a web browser on their PC, phone, or tablet. SaaS uses Openbravo, SuiteCRM, and Acquia in its operations. Openbravo is enterprise resource planning (ERP) software. SuiteCRM is a customer relationship management (CRM) application. In the case of Acquia, it enables the hosting of the Drupal content management system on the Amazon Elastic Compute Cloud (EC2) cloud service to create a digital foundation for delivering web content.

Infrastructure as a Service (IaaS)

IaaS is the most basic category of cloud computing services. In the case of IaaS, IT infrastructure is rented such as servers and virtual machines (VMs), storage, networks, and operating systems from a cloud provider on a pay-as-you-go basis. The IaaS uses CloudStack, OpenStack, and Eucalyptus for its operations. CloudStack supports VMware's vSphere and Xen virtualization and offers a management server with a web dashboard. OpenStack offers modular architecture that provides a component-based way to build clouds. Eucalyptus, an elastic utility computing architecture, is used for linking programs to useful systems.

Serverless Computing

Serverless computing focuses on building application functionality without spending time continually managing the servers and infrastructure required to do so. It overlaps with PaaS. The cloud provider handles the setup, capacity planning, and server management for you. Serverless architectures are highly scalable and event-driven, only using resources when a specific function or trigger occurs.

12.5.3 Key Benefits of Cloud Computing

The idea of cloud computing service is completely different from the traditional ways of computing service and in the traditional ways of relating to IT businesses and resources. The cloud computing service has opened favorable opportunities that have made organizations turn to it for their computing service needs. Some of these possible reasons are as follows:

Security

In cloud computing, security is tighter as a result of the broad set of policies, technologies, and controls. Many of the cloud providers offer the highest security protection to their organizations, thereby protecting data, apps, and infrastructure from potential threats.

Global Scale

The benefits of cloud computing services include the ability to scale elastically. The cloud delivers the right amount of IT resources such as computing power when needed, storage, the right bandwidth when it is needed, and from the right geographic location.

Cost

The capital expenses of buying hardware and software and setting up and running on-site data centers like the racks of servers and the round-the-clock electricity for power and cooling are all eliminated by cloud computing using the IT experts managing the infrastructure.

Performance

The biggest cloud computing services run on a worldwide network of secure data centers, which are regularly upgraded to the latest generation of fast and efficient computing hardware. This offers several benefits over a single corporate data center, including reduced network latency for applications and greater economies of scale.

Speed

Most cloud computing services are provided self-service and on demand. Large amounts of computing resources can be provisioned in minutes. This is typically done with just a few mouse clicks, it gives businesses lots of flexibility, and it takes the pressure off the capacity planning.

Productivity

On-site data centers typically require a lot of "racking and stacking" hardware setup, software patching, and other time-consuming IT management chores. Cloud computing removes the need for many of these tasks, so IT teams can spend time on achieving more important business goals.

12.6   Fog Computing

Just as we have cloud computing, we also have the fog computing. They both complement each other with their respective functionalities.

Fog computing or fog networking, also known as fogging, is an architecture that uses edge devices to carry out a substantial amount of computation, storage, and communication locally and is routed over the Internet backbone

Fog computing can also be defined as the process of extending cloud computing and the various services associated with it to the edge of the network.

The function of the fog computing is to extend the cloud computing to be closer to the things that produce and act on IoT data. In industry, fog computing is also called edge computing. It has the capability to compute and provide storage, application services, and data to end users. The fog nodes also known as devices can be put with a network connection anywhere on the top of a power pole, on a factory floor, on an oil rig, in a vehicle, and even along a railway track. The switches, embedded servers, routers, video surveillance cameras, and industrial controllers are some exam-

works. Figure 12.12 shows a typical fog computing architecture. In FogMNW, it is necessary to have a systematic management of communication and computing resources. The motivation is that with the introduction of fog computing, new mobile services are brought which demand high computing capability. However, the computing task can be split and offloaded to different fog nodes in mobile networks, where they are jointly optimized with communication resources such as bandwidth and power.

Fog computing enhances the serving capability of mobile networks to support advanced applications such as augmented reality/virtual reality (AR/VR). With the computing and storage capabilities introduced by fog computing, future FogMNW like 5G will be enabled to adopt new techniques. Examples of such new techniques are wireless big data that can be stored in nodes and intelligent algorithms like deep learning; all could be supported.

12.7   Edge Computing

A distributed computing idea in which computation is largely or completely performed on distributed device nodes known as smart devices or edge devices as opposed to primarily taking place in a centralized cloud environment is known as edge computing.

ples of fog nodes. The fact that IoT data are analyzed close to where they are collected helps to minimize latency. It not only keeps sensitive data inside the network but also offloads gigabytes of the network traffic from the core network.

The combination of communication and computing resources has made it possible for fog computing-enabled mobile communication networks (FogMNW) to be enabled to achieve much higher capacity than conventional communication net-

The emergence of edge computing was as a result of the computational needs of mobile computing, Internet of Things (IoT), and smart home—smart light, smart TV, etc., and robot vacuum. Connecting directly to cloud may not just be enough. There has to be some augmented reality (AR). This emerging technology called edge computing offers manifold performance improvements for applications with low latency and high bandwidth requirements. The edge computing uses three levels of

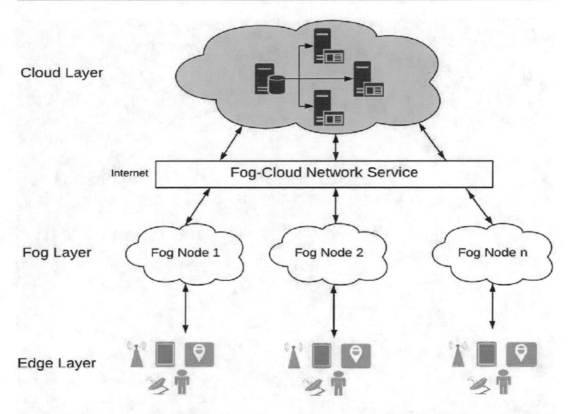

Fig. 12.12 Fog computing architecture. (Source: https://www.semanticscholar.org/paper/Fog-Computing-Architecture%2C-Applications-and-A-Neware/9a7a30830125d34dd85b700f6349e8e19dc995bf)

architecture to accomplish its computing tasks. The first level is the unmodified cloud infrastructure. The second level is the dispersed elements called cloudlets with state cached from the first level, and the third level is the mobile device or IoT device. Fig. 12.13 shows the application architectural diagram of the edge computing structure, while Fig. 12.14 shows the illustration of the edge computing devices, nodes, and cloud.

The edge computing idea has gained wide popularity to the extent that several standardization organizations such as Institute of Electrical and Electronic Engineers (IEEE) and the European Telecommunication Standard Institute (ETSI) are now proposing establishing architectural 5G Working Group and architectural framework called Multi-access Edge Computing (MEC), for third-party application service providers (ASPs), respectively.

Because of edge computing and the interest, it has generated in distributed, low latency, and

reliable machine learning, the tendency now exists in departing from cloud-based and central training and inference toward a novel system called edge machine learning. In this edge system, training data are unevenly distributed over a large number of edge devices such as network base stations and mobile stations including cameras, phones, drones, and vehicles. The edge devices have access to a tiny fraction of the data and training, and inferences are carried out collectively.

12.8 Emerging 5G Networks

There has been a complete evolution of mobile networks from 1G network to now emerging 5G network. The 5G network is an end-to-end ecosystem to enable a fully mobile and connected society. It provides at least a 10-fold improvement in user experience compared to 4G in terms

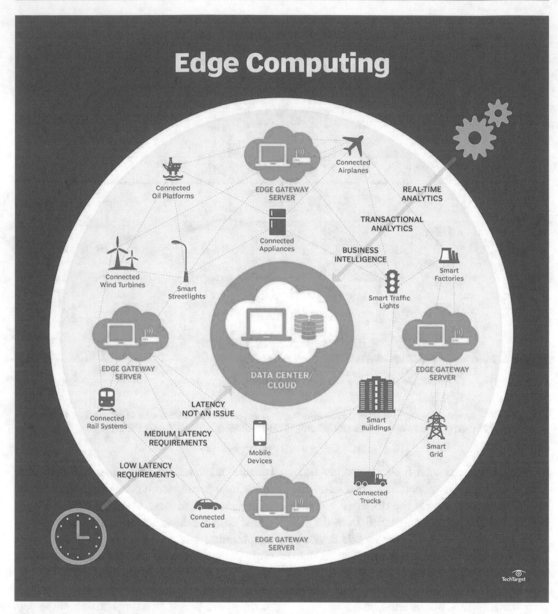

Fig. 12.13 Edge computing application architecture. (Source: https://cdn.ttgtmedia.com/rms/onlineImages/edgecomputing.jpg)

of peak data rates and minimal latency. It supports multi-tenancy and network resource slicing models. The 5G is designed to be a sustainable and scalable technology. As shown in Fig. 12.15, the 5G networks are expected to not just be an evolution from 4G but a full new mobile system.

The 5G design is expected to accomplish the following:

- Connect to everyone
- Connect to objects
- Be environmentally friendly

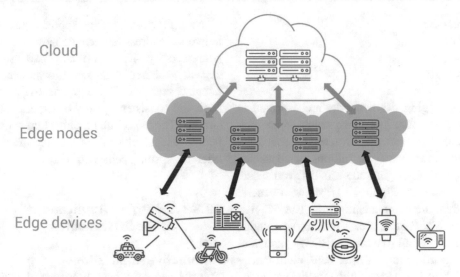

Fig. 12.14 Illustration of edge computing devices, nodes, and cloud. (Source: https://img.alicdn.com/tfs/TB1QdC2KhjaK1RjSZKzXXXVwXXa-3302-1854.png)

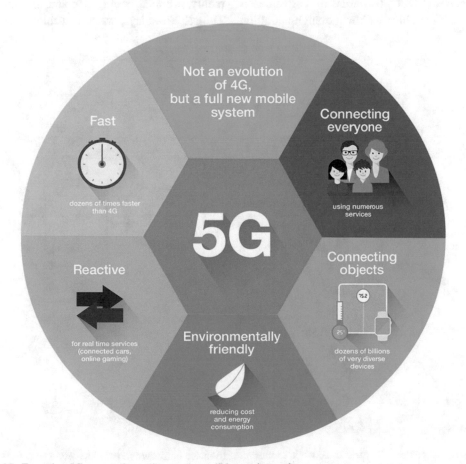

Fig. 12.15 Emerging 5G expectations. (Source: https://blog.aeris.com/)

- Be reactive
- Be fast
- Be flexible
- Be robust

The emerging 5G networks are expected to unleash a massive IoT ecosystems where networks can serve not only in terms of speed but communication needs for billions of connected devices, with the right trade-offs between speed, latency, and cost. Figure 12.16 shows a general 5G cellular network architecture. The 5G networks have advanced access technologies called beam division multiple access (BDMA) and non- and quasi-orthogonal or filter bank multi carrier (FBMC) multiple access unlike the 4G networks. The expected data rate is 10–50 Gbps with an expected frequency band of 30–300 GHz. The bandwidth is 60 GHz with a low density parity check codes (LDPC) forward error correction. It uses packet switching with an ultrahigh-definition video plus virtual reality applications. The 5G networks address six challenges that are not addressed by 4G networks. They are higher capacity, higher data rate, lower end-to-end latency, massive device connectivity, reduced cost, and consistent quality of experience provisioning. The 5G cellular architecture is heterogeneous and included in it are macrocells, microcells, small cells, and relays.

12.8.1 Emerging Applications of the 5G Networks

Automotive Applications

The 5G networks have promises of connecting everything around us to a network that offers the speed, responsiveness, and reach to unlock the full capabilities of technologies such as virtual reality, artificial intelligence and the Internet of Things. Nowadays, we have vehicles like Tesla

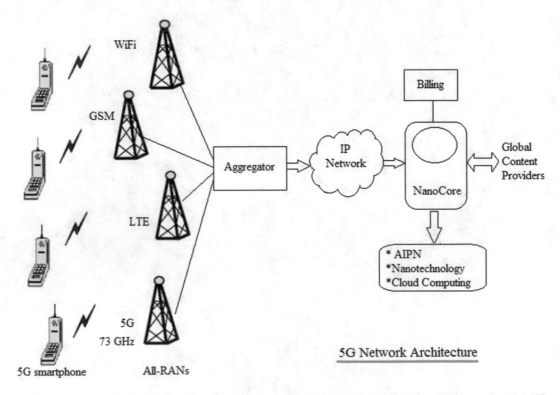

Fig. 12.16 A general 5G cellular network architecture. (Source: https://www.rfwireless-world.com/Tutorials/5G-network-architecture.html)

who offer an autopilot mode that allows drivers to take a quick break, checking in with you periodically to ensure safe parameters. Autonomous vehicles have come a long way but are still in the experimental stage. 4G networks do not offer the speed and responsiveness needed to fully operate self-driving vehicles. An autonomous car operated by Uber killed a pedestrian in Arizona. The question might arise, "How did a self-driving car kill a pedestrian?"

As seen in Fig. 12.17, the operating car was a Volvo XC90 sport utility vehicle outfitted with a sensor system and was in autonomous mode when it struck a pedestrian. There was a human safety driver at the wheel, and the vehicle was going about 40 mph when it struck the pedestrian who was walking her bicycle across the street. Camera recordings from both the interior and exterior of the car show the moments leading to the tragedy but clearly show the safety driver was distracted looking down and her hands were not hovering the steering wheel. Had the safety driver been more alert, this accident could have been prevented. The 5G networks are to be equipped with the responsiveness and the speed needed to ensure fatalities like these are avoided.

Health Care/E-Health

In healthcare, using a 5G network means more efficient, reliable, and accessible way to treat patients. The 5G network will help improve the way we receive and give healthcare significantly. Doctors will implement real-time remote monitoring to track patients in real time allowing them to more effectively treat the patient. File transmission will also be another benefit. MRI files must be sent to a specialist for review. We forget how big these files are and sometimes can take up to 23 h to transmit them on a 4G network. Longer transmission times mean a longer wait for treatment. This also means practitioners can see fewer patients in the same amount of time.

The 5G high-speed network will quickly and reliably transport huge data files of medical

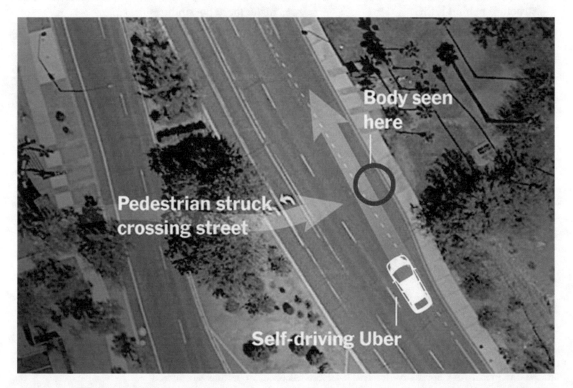

Fig. 12.17 Uber accident diagram. (Source: https://www.nytimes.com/2018/03/19/technology/uber-driverless-fatality.html)

imagery. It is also a more effective way to translate to patients who speak other languages through video conferencing. Artificial intelligence will be able to detect potential diagnosis in 5G networks. The faster a patient can be diagnosed, the faster that patient can be cared for. Figure 12.18 shows the general idea of 5G networks implemented in healthcare.

Virtual Ventures

Virtual reality (VR) offers customers a real lifelike experience to play, learn, and communicate. VR and augmented reality (AR) will be able to offer these lifelike experiences solely because the 5G network supports these claims:

- Reduced latency which implies better interactivity and expanded use cases
- Massive connectivity capacity
- High mobility
- Consistent quality
- Anywhere use case
- Technology developed can support the infrastructure behind it (5G)

Because of these promises being able to attend a sold-out concert from the comfort of your living room, touring homes, virtual surgery, and the experience to visualize products in more different settings will be possible.

Smart Homes

Competitors such as Google and Amazon are pushing for central hubs that control the activity of your home with vocalized commands. 5G connected homes offer:

- Enhanced home entertainment
- Ability to control number amounts of connected devices
- Improvements to technology and services
- Ability to operate without getting up

With these new possibilities, it will allow for a new experience at home. Eventually, smart connected homes will be able to be operated from outside of your home from the comfort of your smartphone anywhere in the world. 5G networks make these possibilities possible.

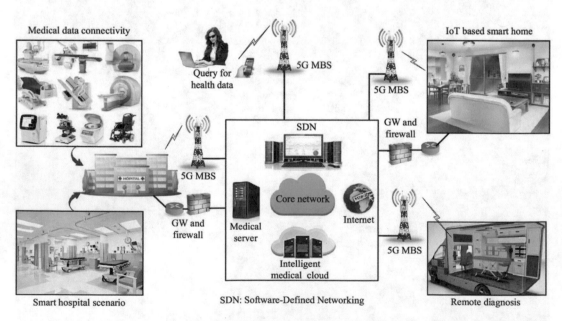

Fig. 12.18 eHealth in 5G. (Source: https://www.semanticscholar.org/paper/A-New-5G-eHealth-Architecture-Based-on-Optical-An-Chowdhury-Hossan/c384ef9cce9e9cedca25f60e28fa92e0cfaf4a31

12.9 Cybersecurity

Cybersecurity is the protection of valuable resources which may include information systems against unauthorized access to or modification of information, whether in storage, processing, or transit, and against the denial of service to authorized users, including those measures necessary to detect, document, and counter such threats.

In the field of information technology, cybersecurity plays a very significant role. This is because users of information and those that send information would like the information to be secured as it goes from sender to receiver without any breach of security. As the use of Internet increases daily as the fastest growing technological infrastructure, so is the increase in cybercrimes. However, due to these emerging technologies, it has become almost impossible to secure our private information in a very effective manner due to increase in cybercrimes.

12.9.1 Cybersecurity Techniques

In the cybersecurity world, the idea of user name and password has been the first measures to protecting information. However, there are other techniques such as the authentication of data, malware scanners, firewalls, and antivirus software.

- Authentication of data—This is where the data sent are checked to make sure they are coming from a trusted and reliable source. Antivirus software is normally used in the authentication process.
- Malware scanners—In this case, a malware software is normally used to scan through all the information and documents that are present in the system to make sure there are no malicious codes or harmful viruses present.
- Firewall—This is a piece of software program that helps to screen out hackers, viruses, and worms that may try to reach the user's computer or network over the Internet.

- Antivirus software—In this case, you have an antivirus software which is a computer program that detects, prevents, and takes action to remove malicious software programs such as viruses and worms.

12.9.2 Cybercrimes

The act of getting information illegally or disturbing the passage of information using a computer as the primary means of commission and theft is called cybercrime.

The list of cybercrimes includes network intrusions and the dissemination of computer viruses, as well as computer-based variations of existing crimes such as identity theft, stalking, bullying, and terrorism which has become a major issue for people and nations. Other possible types of cybercrimes include the following:

- Denial of service (DOS) attacks—This takes place when the attacker explicitly attempts to deny service to the intended user. As part of the DOS is the case of spamming a computer resource such as web server or memory until the server overloads and crashes, causing it to slow down to the extent that the user cannot use the machine.
- Probing—This is a type of hacking where the attacker gains access to the system by knowing the weak point in the system.
- User to root (U2R) attack—This is where the attacker attempts to get the access rights from the host and is able to gain root access to the machine.
- Remote to local (R2L) attack—This is where the perpetrator sends particular packets to a machine over a network in an attempt to gain the local access of the network.

12.9.3 Securing Enterprise Networks

Securing information and enterprise networks can be considered under two broad topics: intrusion detection systems (IDSs) and intrusion prevention systems (IPSs). It has been proven that the right combination of selected algorithms/techniques under both topics always produces better security for a given network. This approach leads to using layers of physical, administrative, electronic, and encrypted systems to protect valuable resources.

Self-similarity properties of networks and auto-reclosing techniques used on long rural power lines and multi-resolution techniques can be used in the development of these IDSs. The need to use quantitative methods to detect intrusion is increasing due to the high false positive and false negative rates of existing IDSs and IPSs. Most information and network security techniques employed by the IDSs and IPSs depend mainly on packet behavior for detection. Both real-time and off-line IDS predictions are applied under the analysis and response stages. The basic IDS architecture involves both centralized and distributed/heterogeneous architecture to ensure effective detection. Proactive responses and corrective responses are also employed. The cybersecurity systems, which are made up of both IDS and IPS, are less expensive to implement compared to existing ones.

Summary

1. Internet of Things (IoT) is the interconnection of smart objects or things. When interconnected to people, it becomes Internet of People (IoP), and when interconnected to service, it becomes Internet of Service (IoS).
2. The advancements in the Internet technologies and connected smart objects have contributed in making IoT a widely pervasive computing.
3. IoT is evolving with "anything" and "everything" being connected to the Internet as well as becoming smarter with enabling of technologies.
4. In IoT infrastructures and services, there are various enabling technologies and cloud-enabled platforms that help to make things happen.
5. The term big data refers to the immense amount of data that is collected every second of every minute of every hour of every day. The collection of such data is always a continuous process. The amount of the data is hard to store and hard to analyze.
6. In building a bridge to a smart city, you have to make sure that everything including all services, all data, and all devices can be seamlessly connected. Every smart city starts with its network.
7. The current server-client model for the

Steganography is the science of hiding current data in another transmission medium.

Steganography is another information security technique used in cybersecurity-related issues. Biometric steganography is a feature that is used to implement steganography in skin tone region of pictures. The method embeds secret data in the skin portion of images of people. Skin detection is then performed in cover images and secret data embedding performed in discrete wavelet transform (DWT) domain because it gives a better performance. Images are the most popular cover objects for steganography. This technique has special applications in the military arena.

Internet of things is not fit for the healthcare industry due to high risk associated with a single point of failure. A distributed or peer-to-peer model for the Internet of things based on blockchain will solve the problem of single point of failure and provide trusted protection against distributed denial of service (DDoS) attacks that use vulnerable IoT devices as botnets.

8. Blockchain has the potentials and economic benefits of solving the problem of scale in the healthcare industry. The sensitive that

medical data are expected to have can be guaranteed by the privacy and transparency that can be offered by using blockchain technology.

9. The different types are private, public, and hybrid cloud computing services.
10. The function of the fog computing is to extend the cloud computing to be closer to the things that produce and act on IoT data.
11. The 5G network is an end-to-end ecosystem to enable a fully mobile and connected society. It provides at least a 10-fold improvement in user experience compared to 4G in terms of peak data rates and minimal latency.
12. In cybersecurity systems, using the right combination of IDS- and IPS-selected algorithms/techniques always produces better security for a given network.

Review Questions

12.1 The advancements in the Internet technologies and connected smart objects have contributed in making IoT a widely pervasive concept.
(a) True
(b) False
12.2 Security is not one of the challenges associated with IoT.
(a) True
(b) False
12.3 The basic layers of IoT architectures are:
(a) Object/perception layer and middleware layer(s)
(b) Object/perception layer and application layer
(c) Middleware layer(s) and application layer
(d) Object/perception layer, middleware layer(s), and application layer
12.4 Connected smart homes are one of the examples of IoT.
(a) True
(b) False
12.5 Blockchain is impacting numerous fields of endeavors such as medicine, manufacturing, and supply chain to name a few with several use cases.
(a) True
(b) False

12.6 Some of the drawbacks of big data are:
(a) Variety and velocity
(b) Velocity and volume
(c) Variety, velocity, and volume
(d) None of the above
12.7 The areas of services associated with cloud computing are:
(a) Platform as a service (PaaS) and software as a service (SaaS).
(b) Infrastructure as a service (IaaS) and serverless
(c) Software as a service (SaaS) and infrastructure as a service (IaaS)
(d) Platform as a service (PaaS), software as a service (SaaS), infrastructure as a service (IaaS), and serverless
(e) None of the above
12.8 Global scale is one of the advantages of cloud computing.
(a) True
(b) False
12.9 In millisecond response time, fog computing runs IoT-enabled applications for real-time control and analytics.
(a) True
(b) False
12.10 The 5G is not designed to be a sustainable and scalable technology.
(a) True
(b) False

Answer: 12.1a, 12.2b, 12.3d, 12.4a, 12.5a, 12.6c, 12.7d, 12.8a, 12.9a, 12.10b

Problems

12.1 What are some of the challenges associated with IoT?
12.2. As IoT is evolving with "anything" and "everything" being connected to the Internet, what other enabling technologies can support it to make it smarter?
12.3 Big data is collected from a variety of different sources, and it comes in three different types. What are those types?
12.4 What is the definition of a smart city? What are the elements that must be present as part of a smart city?

12.5 List and describe some of the smart technologies you can find in a smart farm.

12.6 What are the differences between the traditional and smart cities?

12.7 What are the necessary preparations to becoming a smart city?

12.8. What is a blockchain? What other features does a blockchain have?

12.9 What are some of the blockchain technology applications in the field of medicine?

12.10 (a) What is cloud computing? (b) Name and describe the different types of cloud computing.

12.11 Name and describe the four service areas associated with cloud computing?

12.12 What are the key benefits of the cloud computing?

12.13 (a) What is fog computing? (b) Under what conditions are fog computing maximized and mostly needed?

12.14 What are some of the advantages of cloud computing?

12.15 What are the differences between cloud computing and fog computing?

12.16 What are some of the expected accomplishments of the 5G design?

12.17 Name and describe briefly four of the cybersecurity techniques.

12.18 Name and describe briefly at least four of the cybersecurity crimes.

Appendices

Appendix A: Old Technologies

This appendix covers X.25, frame relay, ISDN, BISDN, ATM, and MPLS. Although these technologies are old, they are still referred to by computing networking engineers. These are packet-switching technologies and X.25 is the oldest. In this appendix, we present a brief discussion of these technologies.

X.25

X.25 has been around since 1976 when the transmission lines were non-digital. It is well debugged and stable. It is in a widespread use. It was the first public data, connection-oriented network. It is packet-switching protocol recommended by CCITT (now ITU) and designed to protect data in transit over low-speed analog transmission facilities. The medium is non-digital, error-prone, and unreliable. It is meant to support relatively unintelligent devices through error-prone networks. Being a virtual circuit protocol, there is a call set-up phase, a data transfer phase, and a call clear phase. The user may have permanent virtual circuits (PVCs) or switched virtual circuits (SVCs).

To ensure reliable communication, X.25 incorporates network address, error detection, error recovery, and flow-control mechanism. Flow-control limits the amount of traffic from the users in order to prevent congestion. X.25 has no routing algorithm. Due to its rich features, X.25 supported several protocols like IP, SNA, and DECnet. It supported the basic security requirements of privacy, integrity, and authentication. In the 1980s, X.25 was replaced by frame relay.

Frame Relay

Frame relay (FR) is a widely accepted, popular, WAN solution designed to provide flexible, high-performance interconnection for regional, national, and international networks. It evolved from narrowband ISDN and X.25 in the mid-1980s. Frame relay can be thought of as the next step after X.25. From X.25, FR borrowed the concept of virtual connections. From ISDN, FR adopted the idea of separation functionality based on control, data, and management planes. It is the packet-switching data service portion of ISDN, which has been offered as a separate service. A frame relay network often consists of switches, frame relay access devices (FRAD), routers, and leased lines.

The key features of a frame relay network include the following:

- Frame relay is basically an interface standard.
- It is a packet-based high-speed technology.
- It is a synchronous protocol that combines the best characteristics of time division multiplexing (TDM) and X.25.
- It places heavy reliance on today's reliable communications infrastructure and eliminates overhead associated with error detection and recovery.
- Frame relay services employ rate enforcement, which determines how traffic is

prioritized and handled when data bursts exceed the committed information rate (CIR).

- Scalability is the hallmark of FR. With ease one can add more bandwidth via the CIR.
- Frame relay network does not guarantee delivery of data and does not acknowledge or retransmit frames.
- Frames are delivered in sequence.

Frame relay takes advantage of the recent advances in transmission medium, such as fiber optics, used in WANs. Optical fiber offers significant advantages such as a substantially lower error rate than that obtained with copper wires. Frame relay was developed on the assumption that the transmission media is reliable and relatively error-free. Therefore, frame relay makes little attempt to detect errors and no attempt whatsoever to correct them. Frame relay combines the best features of time division multiplexing (TDM) used in circuit switching and the statistical time division multiplexing (STDM) used in X.25. It is still in use in some places today.

ISDN and BISDN

For computer networks to be able to simultaneously carry voice, audio, and video in addition to data, the Integrated Service Digital Network (ISDN) and broadband ISDN or BISDN were proposed. ISDN is a digital end-to-end telecommunication wide area network (WAN). It is the first network-based standard for simultaneous integrated voice, data, and video signal transmission over a single pair of wires instead of the separate pairs formerly needed for each service. It enables a single point of access for data, voice, and video. Because it is a digital network, it is able to transmit all forms of data, text, image video, and digitized voice over a single network transparently. This is the feature that makes ISDN very attractive to users.

ISDN is not itself a service but an interface to existing and future services. Two interfaces differing mainly in their carrying capacity are basic access and primary access. The *Basic Rate Interface* (BRI) consists of two full-duplex 64 kbps B channels and one full-duplex 16 kbps

D channel. The total bit rate is 144 kbps. Sometimes one or both of the channels are unused so that we have a B + D interface instead of the 2B + D interface. The basic service serves the needs of most users (residential and commercial).

The higher-capacity *Primary Rate Interface* (PRI) in North America has twenty-three 64 kbps B channels and one 64 kbps D channel (i.e., 23B + D). The total bit rate is 2.048 Mbps. The 23B + D PRI interface uses T1 for the layer one physical transport.

The BRI and PRI offerings of the standard ISDN are collectively referred to as narrowband ISDN (NISDN). Thus NISDN provides services which will be carried by channels based on 64 kbps up to 1.5 Mbps (or 2 Mbps for European standard). The characteristics of NISDN are inadequate for many applications of interest and in meeting the perceived users' needs for higher speed, broader bandwidth, and more flexibility such as video distribution, HDTV, and HiFi stereo. The needs are accommodated in broadband ISDN (BISDN). Consequently, as far as data networks are concerned, real excitement of ISDN comes about when one considers the capabilities of BISDN.

BISDN is regarded as an all-purpose digital network in that it will provide an integrated access that will support a wide variety of applications in a flexible and cost-effective manner. Broadband services include high-speed transmission, video distribution, video telephony, video conferencing, telefax, and TV distribution.

ATM

The transfer mode for BISDN must be able to handle both narrowband and broadband rates, handle both continuous and bursty traffic, satisfy delay and/or loss sensitive quality requirements, and meet unforeseen demands. Neither circuit mode nor packet mode is suitable for meeting all these requirements. Asynchronous transfer mode (ATM), which is something between these two modes, has been selected as the target solution of BISDN. It is a specific packet-oriented transfer mode using an asynchronous time division multiplexing technique.

ATM may be regarded as the foundation on which BISDN is to be built. The term "asynchronous" distinguishes the mode from the synchronous transfer mode (STM) which predominates today's switched networks. In ATM networks, information is organized in fixed-size segment (or slots) called *cells* which may appear at irregular intervals. Each cell consists of a header and an information field. The ATM cell format is shown in Fig. A.1. Using the ATM technique, the information stream of each service is packetized and placed in short fixed-length cells. An ATM cell consists of a 40-bit (5 octet) header field and a 384-bit (48 octet) information field which contains user data.

The ATM switch constitutes an important component of the ATM network. A switch may handle several hundred thousand cells per second at each switch port. In its simplest form, a switch has a number of links to receive and transmit cells. The primary function of an ATM switch is to route cells from an input port onto an appropriate output port at an extremely high bit rate. The switch accomplishes this by examining the fields within each cell header in conjunction with the routing table to determine the outward link.

Applications of ATM technologies and services include Internet backbone, internetworking, frame relay-ATM interworking, voice over ATM, video over ATM, multimedia and broadband services, and broadband satellite network.

MPLS

MPLS stands for multiprotocol label switching. In the OSI model, layer two covers protocols like Ethernet and SONET, while layer three covers Internet-wide addressing and routing using IP protocols. MPLS is in between these traditional layers and may be regarded "layer 2.5 networking protocol." The term "multiprotocol" implies that its techniques are applications to any network layer protocol. MPLS is a protocol-independent packet-switched network.

MPLS is an extension to the existing IP architecture. It is a standard by the Internet Engineering Task Force (IETF) designed to enhance the speed, scalability, and service provisioning capabilities in the Internet. It offers a mechanism for packet-oriented traffic engineering. MPLS network works by prefixing packets with an MPLS header, containing one or more labels. In other words, data packets are assigned labels. This allows one to create end-to-end circuits across any type of transport medium, using any protocol.

MPLS can be used for IP or any other network layer protocols. It can be deployed in corporate networks as well as in backbone networks. It supports virtual private network (VPN) services. Traffic engineering is inherently taken care of in MPLS using explicitly routed point-to-point paths.

A variation of MPLS that supports all optical networks and offers different wavelengths for

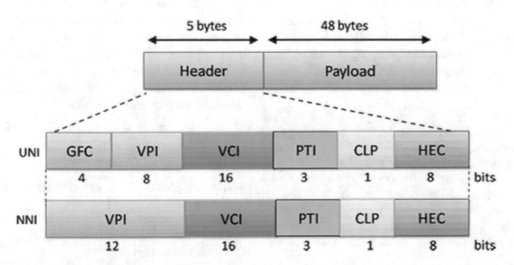

Fig. A.1 ATM cell structure in UNI and NNI formats. (Source: "SONET & SDH optical testing," https://www.gl.com/lightspeed1000-atm-analyzer.html)

Fig. B.1 A typical queueing system

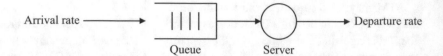

Arrival rate ⟶ | Queue | ⟶ (Server) ⟶ Departure rate

different data flows on the transport layer is the generalized multiprotocol label switching (GMPLS). GMPLS is a set of traffic engineering and optical extensions to existing MPLS routing and signaling protocols. Its goal is cover both the routing and signaling portions of the control plane. It extends MPLS to encompass time division, wavelength, and spatial switching. Although MPLS was designed for packet services, GMPLS is designed to provide a single suite of protocols that would be applicable to all kinds of traffic.

Appendix B: Queueing Theory

Reduced to its most basic form, a computer network consists of communication channels and processors (or nodes). As packets flow from node to node, queues begin to form at different nodes, and they spend a lot of time waiting in queues. For high traffic intensity, the waiting or queueing time can be dominant so that the performance of the network is dictated by the behavior of the queues at the nodes. Analytical derivation of the waiting time requires a knowledge of queueing theory.

This appendix provides a brief introduction to queueing theory. Queueing is simply waiting in lines such as stopping at the toll booth, waiting in line for a bank cashier, stopping at a traffic light, waiting to buy stamps at the post office, and so on. A queue is the waiting line.

> A **queue** consists of a line of people or things waiting to be served and a service center with one or more servers. For example, there would be no need of queueing in a bank if there are infinite number of people serving the customers. But that would be very expensive and impractical.

Queueing theory is applied in several disciplines such as computer systems, traffic management, operations, production, and manufacturing. It plays a significant role in modeling computer communication networks. Since the mid-1960s, performance evaluation of computer communication systems is usually made using queueing models.

Kendall's Notation

In view of the complexity of a data network, we first examine the properties of a single queue. The results from a single-queue model can be extended to model a network of queues. A single queue is comprised of one or more servers and customers waiting for service. As shown in Fig. B.1, the queue is characterized by three quantities:

- The input process
- The service mechanism
- The queue discipline

The *input process* is expressed in terms of the probability distribution of the interarrival times of arriving customers. The *service mechanism* describes the statistical properties of the service process. The *queue discipline* is the rule used to determine how the customers waiting get served. To avoid ambiguity in specifying these characteristics, a queue is usually described in terms of a well-known shorthand notation devised by D. G. Kendall. In Kendall's notation, a queue is characterized by six parameters as follows:

$$A / B / C / K / m / z \qquad (B.1)$$

where the letters denote:

A: Arrival process, i.e., the interarrival time distribution
B: Service process, i.e., the service time distribution
C: Number of servers

K: Maximum capacity of the queue (default = ∞)
m: Population of customers (default = ∞)
z: Service discipline (default = FIFO)

The letters A and B represent the arrival and service processes and assume the following specific letters depending on which probability distribution law is adopted:

D: Constant (deterministic) law, i.e., interarrival/service times are fixed
M: Markov or exponential law, i.e., interarrival/service times are exponentially distributed
G: General law, i.e., nothing is known about the interarrival/service time distribution
GI: General independent law, i.e., all interarrival/service times are independent
E_k: Erlang's law of order k
Hk: Hyperexponential law of order k

The most commonly used service disciplines are:

FIFO: First-in first-out
FCFS: First-come first-serve
LIFO: Last-in first-out
FIRO: First-in random-out

It is common in practice to represent a queue by specifying only the first three symbols of Kendall's notation. In this case, it is assumed that K = ∞, m = ∞, and z = FIFO. Thus, for example, the notation M/M/1 represents a queue in which arrival times are exponentially distributed, service times are exponentially distributed, there is one server, the queue length is infinite, the customer population is infinite, and the service discipline is FIFO. In the same way, an M/G/n queue is one with Poisson arrivals, general service distribution, and n servers.

Example B.1

A single-queue system is denoted by M/G/4/10/200/FCFS. Explain what the operation of the system is.

Solution

The system can be described as follows:

1. The interval arrival times are exponentially distributed.
2. The services times follow a general probability distribution.
3. There are four servers.
4. The buffer size of the queue is 10.
5. The population of customers to be served is 200, i.e., only 200 customers can occupy this queue.
6. The service discipline is first come, first served.

Little's Theorem

To obtain the waiting or queueing time, we apply a useful result, known as *Little's theorem*, after the author of the first formal proof in 1961. The theorem relates the mean number of customers in a queue to the mean arrival rate and the mean waiting time. It states that a queueing system, with average arrival rate λ and mean waiting time per customer $E(W)$, has a mean number of customers in the queue (or average queue length) $E(N_q)$ given by

$$E\left(N_q\right) = \lambda E\left(W\right) \qquad (B.2)$$

The theorem is very general and applies to all kinds of queueing systems. It assumes that the system is in statistical equilibrium or steady state, meaning that the probabilities of the system being in a particular state have settled down and are not changing with time.

It should be noted that Eq. (B.2) is valid irrespective of the operating policies of the queueing system. For example, it holds for an arbitrary network of queues and serves. It also applies to a single queue, excluding the server.

The theorem can be proved in many ways. One of them, the graphical proof, will be given

Fig. B.2 Plot of arrival time and departure time

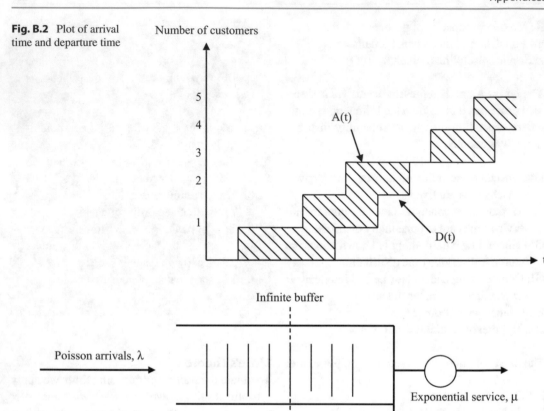

Fig. B.3 M/M/1 queue

here. Suppose we keep track of arrival and departure times of individual customers for a long time t_o. If t_o is large, the number of arrivals would approximately equal to the number of departures. If this number is N_a, then

$$\text{Arrival Rate} = \lambda = \frac{N_a}{t_o} \qquad \text{(B.3)}$$

Let $A(t)$ and $D(t)$ be respectively the number of arrivals and departures in the interval $(0, t_o)$. Figure B.2 shows $A(t)$ and $D(t)$. If we subtract the departure curve from the arrival curve at each time instant, we get the number of customers in the system at that moment. The hatched area in Fig. B.2 represents the total time spent inside the system by all customers. If this is represented by J,

$$\text{Mean time spent in system} = T = \frac{J}{N_a} \qquad \text{(B.4)}$$

From Eqs. (B.3) and (B.4),

Mean number of customers in the system

$$= N = \frac{J}{t_o} = \frac{N_a}{t_o} \times \frac{J}{N_a} \qquad \text{(B.5)}$$

or

$$\boxed{N = \lambda T} \qquad \text{(B.6)}$$

which is Little's theorem.

M/M/1 Queue

Consider the M/M/1 queue shown in Fig. B.3. This is the simplest nontrivial queueing system. It consists of a single-server system with infinite queue size, Poisson arrival process with arrival rate λ, and exponentially distributed service times with service rate μ. The Poisson process has the unique property that future increments are independent of the past history. The queue discipline is FCFS.

The probability of k arrivals in a time interval t is given by the Poisson distribution:

$$p(k) = \frac{(\lambda t)^k}{k!} e^{-\lambda t}, \quad k = 0, 1, 2, \cdots \quad (B.7)$$

(Note that the Poisson arrival process has exponential arrival times.) It is readily shown that the mean or expected value and variance are given by

$$E(k) = \sum_{k=0}^{\infty} kp(k) = \lambda t \quad (B.8a)$$

$$\mathrm{Var}(k) = E\left[(k - E(k))^2\right] = \lambda t \quad (B.8b)$$

One way of analyzing such a queue is to consider its state diagram in Fig. B.4. We say that the system is in state n where there are n customers in the system (in the queue and the server). Notice from Fig. B.4 that λ is the rate of moving from state n to $n+1$ due to an arrival in the system, whereas μ is the rate of moving from state n to $n-1$ due to departure when service is completed. If $N(t)$ is the number of customers in the system (in the queue and the server) at time t, the probability of the queue being in state n at steady state is given by

$$p_n = \lim_{t \to \infty} \mathrm{Prob}\left[N(t) = n\right], \quad n = 0, 1, 2, \cdots \quad (B.9)$$

Our goal is to find p_n and use it to find some performance measures of interest.

Consider when the system is in state 0. Due to an arrival, the rate at which the process leaves state 0 for state 1 is λp_o. Due to a departure, the rate at which the process leaves state 1 for state 0

is μp_1. In order for stationary probability to exist, the rate of leaving state 0 must equal the rate of entering it. Thus

$$\lambda p_o = \mu p_1 \quad (B.10)$$

Consider when the system is in state 1. Since p_1 is the proportion of time the system is in state 1, the total rate at which arrival or departure occurs is $\lambda p_1 + \mu p_1$, which is the rate at which the process leaves state 1. Similarly, the total rate at which the process enters state 1 is $\lambda p_0 + \mu p_2$. Applying the rate-equality principle gives

$$\lambda p_1 + \mu p_1 = \lambda p_0 + \mu p_2 \quad (B.11)$$

We proceed in this manner for the general case of the system being in state n and obtain

$$(\lambda + \mu) p_n = \lambda p_{n-1} + \mu p_{n+1}, \quad n \geq 1 \quad (B.12)$$

The right-hand side of this equation denotes the rate of entering state n, while the left-hand side represents the rate of leaving state n. Equations (B.10), (B.11), and (B.12) are called *balance equations*.

We can solve Eq. (B.12) in several ways. An easy way is to write Eq. (B.12) as

$$\begin{aligned} \lambda p_n - \mu p_{n+1} &= \lambda p_{n-1} - \mu p_n \\ &= \lambda p_{n-2} - \mu p_{n-1} \\ &= \lambda p_{n-3} - \mu p_{n-2} \quad (B.13) \\ &\vdots \quad \vdots \\ &= \lambda p_0 - \mu p_1 = 0 \end{aligned}$$

Thus

$$\lambda p_n = \mu p_{n+1} \quad (B.14)$$

or

$$p_{n+1} = \rho p_n, \quad \rho = \lambda / \mu \quad (B.15)$$

If we apply this repeatedly, we get

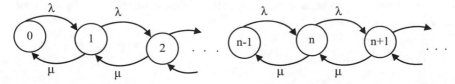

Fig. B.4 State diagram for M/M/1 queue

$$p_{n+1} = r\, p_n = r^2 p_{n-1} = r^3 p_{n-2} = \cdots = r^{n+1} p_0,$$
$$n = 0,1,2,\cdots$$
$$\text{(B.16)}$$

We now apply the probability normalization condition

$$\sum_{n=0}^{\infty} p_n = 1 \qquad \text{(B.17)}$$

and obtain

$$p_0 \left[1 + \sum_{n=1}^{\infty} \rho^n \right] = 1 \qquad \text{(B.18)}$$

If $\rho < 1$, we get

$$p_0 \frac{1}{1-\rho} = 1 \qquad \text{(B.19)}$$

or

$$p_0 = 1 - \rho \qquad \text{(B.20)}$$

From Eqs. (B.15) and (B.20),

$$\boxed{p_n = (1-\rho)\rho^n, \quad n = 1,2,\cdots} \qquad \text{(B.21)}$$

which is a geometric distribution.

Having found p_n, we are now prepared to obtain some performance measures or measures of effectiveness. These include utilization, throughput, the average queue length, and the average service time.

The *utilization U* of the system is the fraction of time that the server is busy. In other words, U is the probability of the server being busy. Thus

$$U = \sum_{n=1}^{\infty} p_n = 1 - p_0 = \rho$$

or

$$\boxed{U = \rho} \qquad \text{(B.22)}$$

The *throughput R* of the system is the rate at which customers leave the queue after service, i.e., the departure rate of the server. Thus

$$\boxed{R = \mu(1-p_0) = \mu\rho = \lambda} \qquad \text{(B.23)}$$

This should be expected because the arrival and departure rates are equal at steady state for the system to be stable.

The average number of customers in the system is

$$E(N) = \sum_{n=0}^{\infty} np_n = \sum_{n=0}^{\infty} n(1-\rho)\rho^n = (1-\rho)\sum_{n=0}^{\infty} n\rho^n$$
$$= (1-\rho)\frac{\rho}{(1-\rho)^2}$$

or

$$\boxed{E(N) = \frac{\rho}{1-\rho}} \qquad \text{(B.24)}$$

Applying Little's formula, we obtain the *average response time* or *average delay* as

$$E(T) = \frac{E(N)}{\lambda} = \frac{1}{\lambda}\frac{\rho}{1-\rho} \qquad \text{(B.25)}$$

Or

$$\boxed{E(T) = \frac{1}{\mu(1-\rho)}} \qquad \text{(B.26)}$$

This is the mean value of the total time spent in the system (i.e., queue and the server).

As shown in Fig. B.5, the average delay $E(T)$ is the sum of the average waiting time $E(W)$ and the average service time $E(S)$, i.e.,

$$E(T) = E(W) + E(S) \qquad \text{(B.27)}$$

Equivalently, the average number of customers $E(N)$ in the system equals the sum of the average of customers waiting $E(N_q)$ in the queue and the average number of customers $E(N_s)$ being served, i.e.,

$$E(N) = E(N_q) + E(N_s) \qquad \text{(B.28)}$$

But the mean service $E(S) = \dfrac{1}{\mu}$. Thus

$$E(W) = E(T) - \frac{1}{\mu} \qquad \text{(B.29)}$$

or

$$\boxed{E(W) = \frac{\rho}{\mu(1-\rho)}} \qquad \text{(B.30)}$$

Fig. B.5 Little's
formula applied to
M/M/1 queue thrice

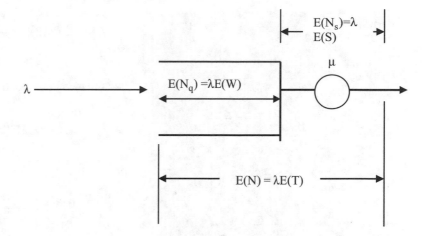

We now apply Little's theorem to find the *average queue length* or the average number of customers waiting in the queue, i.e.,

$$E\left(N_q\right) = \lambda E\left(W\right) = \frac{\rho^2}{1-\rho} \qquad (B.31)$$

Finally, since $E(N) = \lambda E(T)$,

$$E\left(N_s\right) = \lambda E\left(S\right) = \lambda \frac{1}{\mu} = \rho \qquad (B.32)$$

Notice from Eqs. (B.25), (B.31), and (B.32) that the Little's theorem is applied three times. This is also shown in Fig. B.5.

Example B.2

Service at a bank may be modeled as an M/M/1 queue at which customers arrive according to Poisson process. Assume that the mean arrival rate is 1 customer/minute and that the service times are exponentially distributed with mean 40 seconds/customer. (a) Find the average queue length. (b) How long does a customer have to wait in line? (c) Determine the average queue size and the waiting time in the queue if the service time is increased to 50 seconds/customer. (d) Calculate the probability that the number of customers in the system is 3.

Solution

As an M/M/1 queue, we obtain mean arrival rate as
$$\lambda = 1 \text{ customer / minute}$$

and the mean service rate as

$$E\left(S\right) = \frac{1}{m} = 40 \text{ seconds / customer}$$

$$= \frac{40}{60} \text{ minute / customer}$$

Hence, the traffic intensity is

$$\rho = \frac{\lambda}{\mu} = \left(1\right)\left(40/60\right) = \frac{2}{3}$$

(a) The mean queue size is

$$E\left[N_q\right] = \frac{\rho^2}{1-\rho} = \frac{\left(2/3\right)^2}{1-2/3} = 1.333 \text{ customers}$$

(b) The mean waiting time is

$$E\left[W\right] = \frac{r}{m\left(1-r\right)} = \frac{2/3\left(4/6\right)}{\left(1-2/3\right)}$$

$$= 1.333 \text{ minutes}$$

(c) If the mean service time $E(S) = 50$ seconds/customer $= 50/60$ minutes/customer, then

$$\rho = \frac{\lambda}{\mu} = \left(1\right)\left(50/60\right) = \frac{5}{6}$$

$$E\left[N_q\right] = \frac{r^2}{1-r} = \frac{(5/6)^2}{1-5/6}$$
$$= 4.1667 \text{ customers}$$

$$E[W] = \frac{r}{m(1-r)} = \frac{5/6(5/6)}{(1-5/6)}$$
$$= 4.1667 \text{ minutes}$$

We expect the queue size and waiting time to increase if it takes longer time for customers to be served.

(d) For $n = 3$, using Eq. (B.21)

$$p_3 = (1-r)r^3 = (1-2/3)(2/3)^3$$
$$= \frac{8}{81} = 0.09877$$

Other Queueing Systems

Here we briefly consider the M/G/1 and M/D/1 queues. The most general queueing system with single-server and Poisson arrivals is the M/G/1 queue. It is the simplest non-Markovian system. We analyze it assuming that it is in the steady state. An M/G/1 system assumes a FIFO service discipline, an infinite queue size, a Poisson input process (with arrival rate λ), a general service times (with arbitrary but known distribution, mean $\tau = 1/\mu$, and variance σ^2), and one server. To derive the average waiting time of the M/G/1 model requires some effort beyond the scope of this book. We mere state the result:

$$\boxed{E(W) = \frac{\rho\tau}{2(1-\rho)}\left(1+\frac{\sigma^2}{\tau^2}\right)} \quad \text{(B.33)}$$

where $\rho = \lambda/\mu = \lambda\tau$. This is known as *Pollaczek-Khintchine formula* after two Russian mathematicians Pollaczek and Khintchine who derived the formula independently in 1930 and 1932, respectively. The average number of customers $E(N_q)$ in the queue is

$$E\left(N_q\right) = \lambda E(W) = \frac{\rho^2}{2(1-\rho)}\left(1+\frac{\sigma^2}{\tau^2}\right) \quad \text{(B.34)}$$

The average response time is

$$E(T) = E(W) + \tau = \frac{\rho\tau}{2(1-\rho)}\left(1+\frac{\sigma^2}{\tau^2}\right) + \tau \quad \text{(B.35)}$$

and the mean number of customers in the system is

$$E(N) = \lambda E(T) = E(N_q) + \rho \quad \text{(B.36)}$$

or

$$E(N) = \frac{\rho^2}{2(1-\rho)}\left(1+\frac{\sigma^2}{\tau^2}\right) + \rho \quad \text{(B.37)}$$

We may now obtain the mean waiting time for the M/M/1 and M/D/1 queue models as special cases of the M/G/1 model.

For the M/M/1 queue model, a special case of the M/G/1 model, the service times follow an exponential distribution with mean $\tau = 1/\mu$ and variance σ^2. That means,

$$H(t) = \text{Prob}[X \le t] = 1 - e^{-\mu t} \quad \text{(B.38)}$$

Hence,

$$\sigma^2 = \tau^2 \quad \text{(B.39)}$$

Substituting this in Pollaczek-Khintchine formula in Eq. (B.33) gives the mean waiting time as

$$E(W) = \frac{\rho\tau}{(1-\rho)} \quad \text{(B.40)}$$

The M/D/1 queue is another special case of the M/G/1 model. For this model, the service times are constant with the mean value $\tau = 1/\mu$ and variance $\sigma = 0$. Thus Pollaczek-Khintchine formula in Eq. (B.33) gives the mean waiting time as

$$E(W) = \frac{\rho\tau}{2(1-\rho)} \quad \text{(B.41)}$$

It should be noted from Eqs. (B.40) and (B.41) that the waiting time for the M/D/1 model is one-half that for the M/M/1 model, i.e.,

$$E(W)_{M/D/1} = \frac{\rho\tau}{2(1-\rho)} = \frac{1}{2}E(W)_{M/M/1} \quad (B.42)$$

Problems

Example B.3

In the M/G/1 system, prove that:

(a) Prob (the system is empty) = $1-\rho$
(b) Average length of time between busy periods = $1/\lambda$
(c) Average no. of customers served in a busy period = $\dfrac{1}{1-\rho}$

      where $\rho - \lambda\bar{X}$ and $\bar{X}$ is the mean service time.

Solution

(a) Let p_b = Prob. that the system is busy. Then p_b is the fraction of time that the server is busy. At steady state,

$$\text{Arrival rate} = \text{departure rate}$$

$$\lambda = p_b\mu$$

or

$$p_b = \frac{\lambda}{\mu} = \rho$$

      The probability that the system is empty is

$$p_e = 1 - p_b = 1 - \rho$$

(b) The server is busy only when there are arrivals. Hence the average length of time between busy periods = average interarrival rate = $1/\lambda$.

      Alternatively, we recall that if t is the interarrival time,

$$f(t) = \lambda e^{-\lambda t}$$

Hence $E(t) = 1/\lambda$.

(c) Let $E(B)$ = average busy period and $E(I)$ = average idle period. From part (a),

$$p_b = \rho = \frac{E(B)}{E(B) + E(I)}$$

From part (b),

$$E(I) = \text{average length of time}$$
$$\text{between busy periods} = 1/l$$

Hence

$$\rho = \frac{E(B)}{E(B) + \dfrac{1}{\lambda}}$$

Solving for $E(B)$ yields

$$E(B) = \frac{\rho}{\lambda(1-\rho)} = \frac{\bar{X}}{1-\rho}$$

as required.

      The average no. of customers served in a busy period is

$$N_b = \frac{\text{Average length of busy period}}{\text{Average service time}}$$

Hence

$$N_b = \frac{E(B)}{\bar{X}} = \frac{1}{1-\rho}$$

B.1 For the M/M/1 system, find: (a) $E(N^2)$, (b) $E(N(N-1))$, and (c) Var(N).

B.2 In an M/M/1 queue, show that the probability that the number of messages waiting in the queue is greater than a certain number m is

$$P(n > m) = \rho^{m+1}$$

B.3 For an M/M/1 model, what effect will doubling λ and μ have on E[N], E[N_q], and E[W]?

B.4 Customers arrive at a post office according to a Poisson process with 20 customers/hour. There is only one clerk on duty. Customers have exponential distribution of service times with mean of 2 minutes. (a) What is the average number of customers in the post office? (b) What is the probability that an arriving customer finds the clerk idle?

B.5 An airline check-in counter at Philadelphia airport can be modeled as an M/M/1 queue. Passengers arrive at the rate of 7.5 customers per hour, and the service takes 6 minutes on the average. (a) Find the probability that there are fewer than four passengers in the system. (b) On the average, how long does each passenger stay in the system? (c) On the average, how many passengers need to wait?

B.6 An observation is made of a group of telephone subscribers. During the 2-hour observation, 40 calls are made with a total conversation time of 90 minutes. Calculate the traffic intensity and call arrival rate assuming M/M/1 system.

B.7 Customers arrive at a bank at the rate of 1/3 customer per minute. If X denotes the number of customers to arrive in the next 9 minutes, calculate the probability that (a) there will be no customers within that period, (b) exactly three customers will arrive in this period, and (c) at least four customers will arrive. Assume this is a Poisson process.

B.8 At a telephone booth, the mean duration of phone conversation is 4 minutes. If no more than 2-minute mean waiting time for the phone can be tolerated, what is the mean rate of the incoming traffic that the phone can support?

B.9 For an M/M/1 queue operating at fixed $\rho = 0.75$, answer the following questions: (a) Calculate the probability that an arriving customer finds the queue empty. (b) What is the average number of messages stored? (c) What is the average number of messages in service? (d) Is there a single time at which this average number is in service?

B.10 At a toll booth, there is only one "bucket" where each driver drops 25 cents. Assuming that cars arrive according to a Poisson probability distribution at rate 2 cars per minute and that each car takes a *fixed* time 15 seconds to service, find (a) the long-run fraction of time that the system is busy, (b) the average waiting time for each car, (c) the average number of waiting cars, and (d) how much money is collected in 2 hours.

Bibliography[1]

Akujuobi, C. M. and M. N. O. Sadiku, *Introduction to Broadband Communication Systems*. Boca Raton, FL: Chapman & Hall/CRC, 2008.

Alagesan, V. and Natarajan, B., Data collection in wireless sensor network through hybrid mac protocol. *International Journal of Advanced Science and Technology*, January 2020.

Alomari, S., Putra, S., and Taghizadeh, A., A comprehensive study of wireless communication technology for the future mobile devices. *European Journal of Scientific Research*, vol. 60, September 2011.

Askelson, K., Extranets: All part of the third wave. AICPA Infotech Update, July/August, 1998.

Black, U., *Physical Layer Interfaces & Protocols*. Los Alamitos, CA: IEEE Computer Society Press, 1996.

Bradford, N., *The Art of Computer Networking*. Upper Saddle River, NJ: Prentice Hall, 6th edition, 2007.

Cajetan, M. A., and M. N. O. Sadiku, *Introduction to Broadband Communication System*. Boca Raton, FL: Chapman & Hall/CRC Press, 2008, p. 165.

Comer, D. E., *Computer Networks and Internets*. Upper Saddle River, NJ: Prentice Hall, 6th edition, 2011.

Comer, D. E., *Internetworking with TCP/IP Volume 1: Principles Protocol, and Architecture and Internets*. Upper Saddle River, NJ: Prentice Hall, 5th edition, 2006.

Dally, W. J. and B. Towles, *Principles and Practices of Interconnection Networks*. San Francisco, CA: Morgan Kaufmann, 2003.

Denise Grayson et al., Analysis of security threats to MPLS virtual private networks. *International Journal of Critical Infrastructure Protection*, vol. 2, 2009, pp. 146–163.

Forouzan, B. A., *Data Communications and Networking*. New York: McGraw-Hill, 5th edition, 2013.

Forouzan, B. A., *Local Area Networks*. New York: McGraw-Hill, 2003.

Fowlet, D., *Virtual Private Networks: Making the Right Connection*. San Francisco, CA: Morgan Kaufmann Publishers, 1999.

Keiser, G., *Local Area Networks*. New York: McGraw-Hill, 2nd edition, 2002.

Kurose, J. F. and K. W. Ross, *Computer Networking: A Top-down Approach*. Boston, MA: Pearson Education, 6th edition, 2013.

McClain, G. R. (ed.), *Handbook of Networking & Connectivity*. Cambridge, MA: AP Professional, 1994.

McDysan, D., *VPN Applications Guide: Real Solutions for Enterprise Networks*. New York: John Wiley & Sons, 2000, p. 173.

Miloszewski, N., How to install VoIP at home. October 2011, https://www.voipsupply.com/blog/voip-insider/how-to-install-voip-at-home/

Newman, R. C., *Broadband Communications*. Upper Saddle River, NJ: Prentice Hall, 2002.

Phaltankar, K. M., *Practical Guide for Implementing Secure Intranets and Extranets*. Norwood, MA: Artech House, 2000.

Rajiv., *Satellite Internet technology and applications*. July 2022, https://www.rfpage.com/satellite-internet-technology-and-applications/

Rizvi, S. A. M. and V. K. Sharma, *An Introduction to Computer Networks*. Oxford, UK: Alpha Science International, 2011.

---

[1] This bibliography includes only books published on or after 1990. These are the best resources for learning more about computer networks.

Robertazzi, T. G., *Introduction to Computer Networking*. New York: Springer, 2017.

Sadiku, M. N. O., *Metropolitan Area Networks*. Boca Raton, FL: CRC Press, 1995.

Sadiku, M. N. O. and M. Ilyas, *Simulation of Local Area Networks*. Boca Raton, FL: CRC Press, 1995.

Sadiku, M. N. O., *Optical and Wireless Communications: Next Generation Networks*. Boca Raton, FL: CRC Press, 2002.

Sadiku, M. N. O. and S. M. Musa, *Computer Communication for Metropolitan and Wire Area Networks*. New York: Springer, 2010.

Sadiku, M. N. O. and S. M. Musa, *Performance Analysis of Computer Networks*. New York: Springer, 2013.

Sagiroglu, S. and Sinanc, D. (2013) Big Data: A Review, Collaboration Technologies and Systems (CTS). 2013 International Conference on Digital Object Identifier, 42-47.

Scott, C., P. Wolfe, and M. Erwin, *Virtual Private Networks*. Sebastopol, CA: O'Reilly & Associates, 1998.

Shay, W. A., *Understanding Data Communications and Networks*. Belmont, CA: Brooks/Cole, 3rd edition, 2009.

Sarkar, N., *Tools for Teaching Computer Networking and hardware concepts*. Information Science Publishing, 2006.

Stallings, W., *Local and Metropolitan Area Networks*. Upper Saddle River, NJ: Prentice Hall, 5th edition, 1997.

Stallings, W., *Data and Computer Communications*. Upper Saddle River, NJ: Prentice Hall, 6th edition, 2002.

Tanebaum, A. S., *Computer Networks*. Upper Saddle River, NJ: Prentice Hall, 5th edition, 2011.

Walrand, J., *Communication Networks: A First Course*. Homewood, IL: Irwin, 1991.

William Stallings., *Data and Computer Communications*. 10th ed., Pub. Pearson, New York, 2014, p. 162.

Wu, C. H. J. and J. D. Irwin, *Introduction to Computer Networks and Cybersecurity*. Boca Raton, FL: CRC Press, 2013.

Yu, S., and Park, Y., *SLUA-WSN: Secure and lightweight three-factor-based user authentication protocol for wireless sensor networks. Sensors*, vol. 20, no. 15, 2020.

Yuan, R. and W. T. Strayer, *Virtual Private Networks*. Boston, MA: Addison-Wesley, 2001.

Zheng, Y. and S. Akhtar, *Networks for Computer Scientists and Engineers*. New York: Oxford University Press, 2002.

Index

© The Editor(s) (if applicable) and The Author(s), under exclusive license to Springer Nature
Switzerland AG 2022
M. N. O. Sadiku, C. M. Akujuobi, *Fundamentals of Computer Networks*,
https://doi.org/10.1007/978-3-031-09417-0

Printed in the United States
by Baker & Taylor Publisher Services